Problem Puss

by the same author

The Cat Maintenance Manual
The Cat Maintenance Log Book

Grace McHattie

Problem Puss

A Methuen Paperback

A Methuen Paperback

First published in Great Britain 1987
by Methuen London Ltd
11 New Fetter Lane, London EC4P 4EE
© 1987 Grace McHattie

Printed in Great Britain
by Richard Clay Ltd, Bungay, Suffolk

British Library Cataloguing in Publication Data

McHattie, Grace
   Problem puss.
   1. Cats – Behavior
   I. Title
   636.8   SF446.5

   ISBN 0–413–15870–5

Contents

Author's Note

Throughout most of this book,
I have referred to cats as 'he'
for two reasons. One: it saves
using the cumbersome 'he or she'
or the impersonal 'it'. And two:
male cats – like males of all
species – probably cause the
most problems!

Introduction

It has been said that there are no bad dogs — only bad owners.

It is also true to say that there are no bad cats — but that doesn't presuppose bad owners. Owners who care about their cats and the problems they may encounter are certainly not bad owners. Rather, they may be owners who don't fully understand their cats and their cats' needs.

Their 'problem pusses' may be scratching the furniture, wetting or spraying where they shouldn't, biting their owners, fighting one another or behaving in any one of a hundred ways which appears to the owner to be anti-social or wrong.

Often the behaviour is a cry for help; understand the problem, aid the cat and the problem will disappear. Sometimes the behaviour is quite natural but misunderstood by the owner.

This book is not only an attempt to help owners to understand their cats better but it also gives information which cannot be found elsewhere on problems and their (often simple) solutions. It explains cats' behaviour and tells you how you can train your cat so that problems never arise.

It is important to understand the natural behaviour of your cat. Not only will it enable you to live peaceably and happily with your cat, but it will also enrich your relationship, simplify your life and help to keep your cat healthy and happy.

The cat is one of the most unique animals on this earth. No other creature combines domesticity and wildness in one body to such a degree. Few creatures are as intelligent as the cat, which can use problem-solving techniques, think creatively and employ insight when confronted with the unfamiliar, as well as possessing a superior memory. The cat suffers from diseases unique to itself; it has different nutritional requirements from any other creature; and it can suffer from stress if its nutritional, emotional and physical requirements are not met.

Cats are often kept as pets because they are 'easier' than other pets. Certainly they are easier in that they can amuse themselves if they have a busy owner; if necessary, they can let themselves in and out of the house via the cat-door when alone and they don't need to be taken for walks. But in some respects they are more 'difficult' to keep than other pets – because of these emotional, physical and intellectual needs. As pets, they take longer to get to know; owners must learn to understand their spoken language, their body language, their likes and dislikes. I hope that this book will help the caring and observant owner to understand his or her cat in a way which will make both their lives happier and more fulfilling.

Sharing your home with a cat can be as interesting – and as difficult – as sharing it with another human being. Yet few of us look for a spouse in the hasty, haphazard way in which we shop for a cat. Then we wonder why we have problems!

I estimate that at least 95 per cent of cat owners have

some sort of problem with their pet. It may be something very simple – such as the cat eating houseplants – or something the owner simply can't live with any more – such as the cat urinating all over the house or even eating the furniture.

Often, the owner thinks he or she is the only one suffering from this problem. They feel confused – even guilty – and don't know where to turn for help.

Vets are often unhelpful or disinterested. Few have studied behavioural problems, especially in the cat. Veterinary associations are now becoming concerned about the lack of this knowledge among their members and are recommending that vets should offer more assistance to those whose pets have behavioural problems. But it will be a long time before this assistance is available to all who need it. In the meantime, I hope this book will help many worried owners.

The cat is now the number one pet in the United Kingdom. There are $6\frac{1}{2}$ million of them and they outnumber dogs by a full million.

Now that they are number one, it is certainly time that we began to understand these intriguing and unusual pets to the extent which they deserve.

Behaviour

Kittening

Development

Reason for kneading

Washing behaviour

Play

Fighting

Hunting

- the cruelty myth

The solitary cat?

The group cat

Hierarchy

Territory

- marking

Creatures of habit

Behaviour

It is important for an owner to have a knowledge of cat behaviour. By knowing how the 'average' cat behaves or reacts to a situation and by observing your own cat in the same situation, you will come to know your cat better. You will notice if his behaviour changes; and this is vital, for sometimes the first symptom of an illness is a change in behaviour – for better or for worse.

Perhaps the best place to begin studying cat behaviour is at birth.

Any cat which is to have kittens – especially a first-time mother – should be carefully observed by the owner as the date of birth approaches (approximately sixty-five days after conception). Owners should always be present at the birth; it may be necessary to open the birth sacs or help to cut the umbilical cord.

In a feral (wild) cat colony, female cats will act as each other's 'midwives' – helping to cut cords and cleaning up new arrivals.

Many cats are very frightened when birth contractions start and they need the presence and reassurance of a calm owner.

A first-time mother may even not recognise her first-born for what it is; one of my cats was terrified when her first kitten started mewling immediately after birth and she rushed to me for reassurance.

By placing the mother back in her bed with her kitten

or kittens at her teats, her mothering instinct begins to take over as the kitten starts to suckle.

Some female cats *do* kill their kittens after birth; often, it is believed, because they have been disturbed or frightened. I believe that some cats may kill their kittens at birth *simply because they don't know what they are*. A new-born, wriggling, squealing kitten must look very like a small rat to a confused cat. This is another good reason for an owner being present.

New-born kittens can't see, hear or walk, and are totally dependent on their mothers for some weeks. In feral or stray litters (and sometimes in breeders' homes when several queens have litters at around the same time), litters become 'interchangeable'. Mothers will look after, and even suckle, kittens other than their own, a behaviour which has high survival value in the wild.

Often, females who have not given birth will become 'babysitters', looking after a litter while the mother goes off to feed or hunt. (Male cats, even neutered male cats, are often not so accepting of kittens. It is vital that, if your cat has kittens, you ensure that there is no way that strange males can get to them – either through doors, windows or cat-flaps. There have been cases where kittens have been taken off and killed by male cats not belonging to the household.) Males, as a rule, take no hand in the rearing of kittens, although some male pedigree cats do help with washing and playtime. As most pedigree cats have been bred to exhibit qualities of friendliness this can't really be considered 'natural' behaviour in that it is unlikely that male members of a feral colony would behave in this way.

A kitten's eyes will open at around ten days and the

eyes will be deep blue. This will change to the permanent colour at about twelve weeks. Kittens will 'swim' around on their tummies, paddling with their legs, until around three weeks, when they will begin to walk shakily. It will be another week or two before they are able to run.

During this time, they are being fed regularly, on demand, by their mother. A kitten will latch onto a teat which it will find by smell and sound (the mother's purring) and then knead on either side of the teat, using a rhythmic, regular movement, with both paws. This helps to express the milk and make it flow more swiftly.

The kneading movement becomes associated in the kitten's mind with a very pleasurable experience (drinking his mother's milk) and, in adult life, when he is happy, he will repeat these kneading movements with his paws to express his happiness. This is why many adult cats will knead their owners' laps before settling down. It is often painful to the owner but it is unlikely that you will ever be able to train your cat out of doing this as it has been associated in his mind, from earliest times, with pleasure. All you can do is to keep his claws trimmed!

When kittens have finished their feed, their mother will clean them. How often she does this is directly related to her own upbringing – the cat who was cleaned frequently by her mother will frequently clean her young, while a cat who received little more than a 'lick and a spit' from her mother will give her own kittens a more cursory wash.

The mother cat will also lick the kittens' bottoms to stimulate excretion of waste products, which she will also lick up. If this sounds somewhat unpleasant, it has high survival value. In the wild, the cat's nest would have to be kept spotlessly clean and free from any smell which might attract a predator to the young. The mother herself will excrete well away from the nest.

A kitten's first comforting experiences will be sucking his mother's milk and being licked all over. So licking becomes, for an adult cat, not only a means of washing, but also a source of relaxation, exerting a calming effect.

Everyone will, at some time, have noticed their cat failing in a task – perhaps falling when trying to jump from one obstacle to another – and immediately sitting down to give himself a few hard licks. This is often put down to 'saving face' or to an embarrassed reaction, but in fact the cat has probably given himself a slight scare, and the licking has a calming effect.

Another reason for this washing may be to clean away the scent of any sweat. Cats only sweat through their paw pads and it is possible for them to perspire after a fright. By cleaning off any sweat, it is less likely that a predator would be able to find them or follow them.

As the kittens get a little older, like all children everywhere, they come to dislike being washed. They will then wriggle and squirm, trying to get away from their mother and her insistent tongue. To stop them wriggling, the mother cat will gently grip her kitten around the neck using her teeth. The kitten will immediately go limp. This is probably a protective device; should the kitten be grasped by the neck by a predator, it would 'play dead'.

This is also the source of the behaviour noticed by many owners (who are often worried about it) when, during boisterous play, cats or kittens will close their teeth around one another's necks. Usually, they are not doing this with enough pressure to hurt the other cat or kitten – it is simply a controlling device as the other combatant is expected to go limp (and usually does), whereupon it receives a vigorous washing from the 'victor'. This behaviour can sometimes be seen when a cat grasps an owner's wrist with his teeth. This is not a 'bite' in the true sense of the word but an indication that that wrist is doing something the cat would rather it did not do. The grasp of the teeth is simply to stop it.

Kittens will begin to play at around the age of one month. Their play involves pouncing on their mother's tail, which she will wave from side to side to attract them. This is, in fact, the first stage in learning to hunt. Cats are attracted to their prey by movement. Their eyesight is not good close to, which explains why a cat will often be unable to find a treat placed directly under his nose. However, should that treat somehow begin to move, the cat will spot it at once and pounce on it.

Kittens will also chase one another, tumbling over and having 'play fights'. This is all training for life out of doors, when they may have to defend themselves against a strange cat.

The favoured fighting position is underneath. Here, the 'top cat' is definitely at a disadvantage. The cat underneath, lying on his back, has got five sets of weapons free; his four sets of claws, and his teeth. He will grasp or smack with his front paws, while scrabbling with his back paws on the underside of his littermate's tummy. The unfortunate cat on top can only bite and perhaps give half-hearted swipes with one paw.

In the wild, the mother cat will begin to teach her kittens to hunt at this age. She will do this by catching a small prey animal, killing it and bringing it back to the nest for the kittens to 'practise' on. They will learn the feel and smell of a prey animal and they will toss it around and bat at it with their paws. Later, the mother cat will bring live prey back to the nest and release it for the kittens to catch and kill.

It is what is seen as 'playing' with prey which makes some people believe that cats are cruel. The cat has lived with this unfair condemnation for centuries.

A cat or kitten will, at times, corner its prey and will proceed to bat it on the head with his paws. This is with no thought of torture (a purely human concept) in his mind but is designed to subdue the prey if the cat is unable to deliver a killing bite immediately (perhaps because the prey is standing in such a way that the cat is unable to grip its neck).

If you can imagine a cat trying to catch a cornered rat, you will realise that the rat will be desperate and will be trying to bite the cat with its sharp teeth. Such a bite could prove fatal to a cat, so the cat paws the rat on the head, which has the immediate effect of forcing the rat to lower its head. It is then unable to leap up to bite the cat. The cat will continue to do this until the rat is manoeuvred into a position where the cat will be able to kill it with a bite. Sometimes, when this is not possible immediately, the cat will hook the prey with his claws and toss it in the air. This is meant to confuse the prey sufficiently to enable the cat to give the killing bite. It is often this which is seen as 'cruelty', especially if the prey is smaller and less dangerous than a rat – but a cat is unable to modify his response to suit the size of his prey. The killing behaviour is also seen as 'cruel' if it continues after the prey is dead. But for a cat to catch anything, he has to be as 'psyched up' as any human athlete, with his adrenalin flowing and his reactions fine-tuned. As any athlete will know, it is impossible simply to switch off this state once the stimulus has ended. So when the hunting stimulus ends with the death of the prey, it is impossible for a cat to switch off his hunting behaviour. It may take some time for him to calm down sufficiently to even eat what he has caught.

It is this behaviour which is so often misunderstood.

The action of a cat catching his prey, and then letting go of it, only to catch it again when it runs away, is also seen as a form of torture or cruelty. But as the cat's hunting instinct is only triggered by moving objects, he may well lose interest in any prey which is temporarily 'playing dead', only to have his interest revived as soon as the prey begins to move.

Cats, whether they have kittens or not, will often bring prey back to their 'nest' or, in the case of house cats, their home. Many people believe they are bringing presents to their owners and, indeed, there may be some idea in the cat's mind that he is contributing to the communal food stores or helping to teach his humans how to hunt. However, most of the time he is probably just looking for a quiet, secure place where he can eat undisturbed. Many cats will growl ferociously if they have brought prey home and an owner or another cat approaches.

A cat will always hunt alone and this has led to the myth that the cat is a 'solitary' creature. Cats are solitary hunters – but they are social animals. They will live, for preference, in groups. Feral cats will join together in groups, for in the group lies greater security. Danger noticed by one will immediately be transmitted to all, and chances of survival are increased.

The boss cats of these feral groups will always be unneutered males – not necessarily the largest male, but the male which is the most aggressive. He will fight with any newcomers to preserve his status, and these fights, although incredibly noisy, involve a great deal of ritual and it is rare for serious injury to result. The boss cat's aim is simply to show who is boss – he has no desire actually to hurt any of the pretenders to his throne. Once he has won his fight with the newcomer, he may even mount the cat (whichever sex it is) in a simulation of copulation. This, too, is simply to demonstrate who is 'boss'.

Fights also occur over a female in season, and tom-cats will congregate from many miles away to pay court. (This explains why many unneutered male house cats disappear for days at a time, often coming back scratched or with a torn ear, even if there are no female cats living nearby.) Squabbling and fighting will take place, not so much for the female's favours, but to lay claim to the territory on which she lives. The female, however, will make her own choice of mate, and this will not necessarily be the winning cat.

The female hierarchy of the feral group is linked more to the number of litters she has produced, a fertile female holding a higher position than one who has had fewer litters. Cats which have been neutered or spayed will hold a lower position in the hierarchy of a feral group.

The higher the status of a cat in the group, the more

food he will get and the more comfortable his sleeping position, as other cats will defer to his wishes. In return, he will protect the group to the best of his ability.

Cats are very supportive towards one another and a call of distress from one will immediately result in the others of the group rushing to his aid. There are many stories of house cats insisting on being let out and, when they return, they are helping home an injured housemate.

In the average house group of cats (those living with an owner) the hierarchy will still exist but may differ substantially from what it would be in the wild.

As many as 90 per cent of house cats are neutered in the south of England and no real problems should occur in a small household where all the cats are neutered. The balance of power may shift, though, and, in a household where both sexes are kept, the 'boss cat' could well be a female, especially if she has had a litter in the past.

Cats are very territorial and will strongly resent interlopers in their territory. In the wild, the size of territory available to them would be closely linked to the amount of food available in that area in the form of prey. Neutered house cats will have much smaller territories than feral or farm cats, as they are usually not dependent on their territory to provide their food.

In urban areas, the territories of house cats will overlap. Each cat will consider his own garden – and possibly those on either side if another cat doesn't live there – as his own and will defend it if it is entered by a strange cat. Towards the edges of his territorial boundaries, other cats will be tolerated as the outer fringes of their territories overlap with his.

A cat will patrol his territory along clearly defined, unchanging paths and he will mark it as his own daily. He will do this by scratching tree trunks with his claws and possibly spraying urine on certain markers such as trees, bushes or buildings, or by rubbing his bottom against these markers. As there are scent glands in the anal area, these glands will mark the chosen areas with the cat's scent.

Other cats, when investigating this area, will sniff at the markers. An assertive cat will spray over the spray markers, literally 'leaving a calling card'. The cat in residence will spray over this again and he will also continue to spray where only he has sprayed, as the smell lessens with time. It is believed that cats can tell not only which cat 'owns' a piece of territory by the smell, but how long it is since he visited it by the strength of the remaining smell.

Even neutered cats, which don't spray inside the home, will mark their outdoor territory in this way. Problems only really arise for the owner when they decide to start marking inside as well!

Defaecation and urination will also play some part in the marking of territory. Although most cats will 'cover up' faeces and urine with soil (or, indoors, with litter) enough scent will remain to act as another territorial marker. In the case of very assertive cats, such as the boss cats of feral colonies, faeces and urine won't be covered up at all. This obviously means that the scent will be much stronger, acting as a strong signal to other cats that the territory belongs to another.

Squabbles over territory often occur when two cats come face to face on their territory pathways. Much ritualistic hissing will take place and an aggressive stance will be adopted which can lead to a fight if one cat does not give way, usually by turning round and going back the way he came.

Perhaps territorial pathways explain the myth of the 'cat who walks alone' – they do like to be alone on their own patch. Cats particularly dislike other cats running up to them and even cats which are usually friendly will react angrily to this.

Above all, cats are creatures of habit. They have their own territory, which they hate to have changed, and their own friends – animal and human – and change is not welcomed here either. Cats will often react in an extraordinary manner to anything different, however minor. For example, if a much-loved and familiar owner wears a hat for the first time, their cat may run away from them in fear. Cats are often also wary of white coats – this isn't necessarily because they associate them with vets – and this can be a problem for any owner who wants to take his or her cat to a show, as judges wear white coats. So, for anyone who wants a show career for their cat, they should get him used from kittenhood not only to being handled by strangers, but to being handled by strangers in white coats.

Language

Vocal language

- types of call
- the talkative cat
- hissing
- the growl
- purring
- chatter

Body language

- the welcoming cat
- tails
- ears
- eyes
- whiskers
- the anxious cat

How to use body language to talk to your cat

Why cats like people who don't like them

'Mad spells'

Language

Whoever coined the phrase 'dumb animals' certainly wasn't talking about cats! They have a large variety of calls, many of which have distinct and different meanings. These calls are made in three ways: with the mouth open, with the mouth closed and with the open mouth closing. Certain calls will be used for a specific purpose – asking for food, for example. A 'please feed me' miaow will not only be understood by the owner, but by any other cats in the house, who will come running when they hear it in case the food is forthcoming.

The house cat will, in fact, 'talk' more than the feral or farm cat, who will have a much smaller vocabulary. The house cat realises that the spoken word is of some importance to its human owners, as those owners will usually reinforce actions (feeding the cat, opening the door, etc.) with spoken language. The cat will then reinforce *his* actions (going to his food bowl, sitting by the door, etc.) with spoken language.

Most owners will recognise many of their cats' phrases such as the trill of welcome, the happy or anxious miaow, the bloodcurdling howl of fighting. Hisses also have some importance in the cat's vocabulary as a warning signal. Hissing will deter the advance of another cat, not just because of the noise, but because the air expelled from the hissing cat's mouth is off-putting when directed in the other cat's face. Cats dislike the hiss so much that they will run from spray cans, bicycle pumps and the noise of air brakes on lorries and buses.

Many cats, but not all, have a very effective growl – the meaning of which is immediately apparent! The possession of a growl appears to be hereditary – many families of cats are great growlers while others seem incapable of making the noise.

Purring displays different degrees of pleasure; the rougher the purr, the happier your cat is feeling. Most people have heard the very light, delicate purr of the cat who is purring just out of politeness! (In the case of cats which have, in the past, suffered a respiratory illness such as cat 'flu, the purr may always sound rough and rasping.)

Purring is not always a sign of contentment, however. A sick cat will purr as a sign that he is in pain and an anxious cat may purr from nerves.

A cat will sometimes 'chatter'. This odd sound is made by chattering the teeth together and is often heard when a cat is indoors, looking out of the window at a bird just out of reach. The cat has actually shaped his mouth into the correct shape to make a killing bite and then makes the chattering sound in frustration. Some cats will also make the killing bite shape with their mouths, the lips drawn back from the teeth, but will make a pathetic mewling sound as if in pain.

Body language, too, is rich and varied in the case of the cat. Everyone has been welcomed home at one time or another by an ecstatic cat, some of whom will come running as soon as they hear their owner walking up the path, or driving their car up to the house (they can distinguish between footsteps of owners and strangers, and pick up the sound of one car out of all the others).

The welcoming cat will often bounce up against his owner, tail and head high, back slightly arched, with his front paws raising off the ground. He will rub his body against his owner's legs, often accompanying this with a welcoming chirrup or trill. He may then rub his face and chin against his owner and will try to reach his owner's face if possible. A cat's scent glands are on chin, lips and anus, so by rubbing his face and chin against you, he is covering you with his scent, letting other cats know that you are 'his'. He may follow this by turning round and sticking his bottom in your face — as this is another location of the scent glands, you can only assume you are being invited to have a sniff! When cats which know each other well meet after an absence, they will first sniff one another's faces, then sniff one another's bottoms.

If a cat does greet you by sniffing your nose and rubbing against your face, just sniff back. Don't try blowing down his nose as you might with a dog or a horse — a cat will take it as an unfriendly act; a silent hiss.

A happy cat will carry his head and ears high and his tail will stick straight up into the air. If he's feeling particularly happy, his tail may curve over towards his back at the tip. A slowly wagging tail will show that he is alert, even though half-closed eyes may belie this. A wildly-thrashing tail will show that he's angry. Mother cats will wave their tails at their young offspring to teach them to chase and pounce. And many cats, when they want to go to sleep, will cover their noses with their tails or a paw, to decrease their oxygen intake and hasten sleep.

Ears are an excellent indicator of a cat's mood. Upright ears mean your cat is alert and happy and he will swivel his ears when extremely happy – while eating, for instance. When the ears swivel back, the cat is contemplating attack, whereas a defensive cat will flatten his ears sideways. If you see two cats having a disagreement, you will be able to tell who is the aggressor and who is being attacked by the position of their ears.

Squabbling cats will stare fixedly at one another and at least one of them will be hunkering down, bouncing on his back legs like an athlete on a starting block, ready to spring. The fixed stare is threatening behaviour. Often, confrontation can be averted simply by placing an obstacle in their line of sight. When they can no longer see one another, many cats will simply not bother to have their fight. An unassertive cat will often sit with his back to other cats (or his owner) to show that he is not a threat to the other. Very few cats will ever attack a cat whose back is turned to them.

Another example of non-aggressive behaviour is blinking. If two cats confront one another and one does not wish to become involved in a fight, he will blink several times to show that he is not posing a threat to the other cat.

Half-closed eyes will show contentment – even love – as when a cat with half-closed eyes will sit purring on his owner's lap. Some owners believe that blinking slowly at their cat is the feline equivalent of giving them a kiss! If the cat slowly closes his eyes in response, the kiss is returned.

The size of pupils is an indication of your cat's mood too. When his pupils enlarge, it indicates that he has become alert and is very interested in what is going on. This may be seen in response to food being provided, or in response to a confrontation with another cat.

Whiskers, the cat's 'sensors', will also move into different positions in response to mood. When fighting, the whiskers will be drawn back and will help to emphasise the snarl, while they will be drawn forward in pleasurable anticipation.

An anxious cat will twitch his ears and lick his lips rapidly. A very worried or nervous cat will begin to 'flehm'. He will begin to gasp and inhale air through his open mouth. The mouth contains an organ which allows a cat to taste and smell at the same time and the worried cat is using this organ literally to taste and smell danger.

On the other hand, yawning is seen as a sign of reassurance in the cat world.

You can use this knowledge of body language to get to know a nervous cat or kitten. In some sad cases, I have been asked how an owner can make friends with a newly purchased kitten which doesn't identify with humans at all. Any cat or kitten needs to have been brought up with people, as the important 'imprinting' (when the kitten learns to know and trust human beings) happens at the earliest stages of development. Some kittens, unfortunately, have been brought up in 'kitten farms' where they are perhaps left in a shed with their mother and rarely see a human being until they are sold. Such kittens will be frightened of their new owners and will spend much of their time hiding. The only answer is great patience combined with acting like a cat! The owner will have to spend some time regularly each day in a room with the kitten. Get down to the kitten's level, by lying on the floor, but not directly facing the kitten. Lie at an angle to him and don't stare at him directly. Blink slowly and, every so often, give a wide yawn. Make no sudden movements or noises, and spend as much time as you can with the kitten in this way. Unfortunately, such a kitten may never be quite as friendly as one brought up in a family environment, although there are exceptions. Some completely wild, feral cats have joined households and, with great patience and perseverance on the part of their owners, have become well-adjusted and friendly house cats.

The behaviour described above (i.e. being in the same room as a cat but not looking at it directly) is considered 'non-aggressive' by a cat and explains why people who say they don't like cats always find that cats go straight to their laps. As they don't like cats, they don't look at them, so cats wrongly interpret this as friendly behaviour! Anyone who doesn't want a cat on his lap should spend a few minutes staring at the offending creature when they first enter the room. (However, 'cat haters' never take this advice. It is frequently a source of pride with them that cats always want to sit on their laps.)

I am sometimes asked about the 'mad spells' most cats will have from time to time when, for no apparent reason, they will suddenly leap up, rush around the house at a great rate, jumping onto furniture and attacking their toys. There may be several reasons for this: either an excess of energy (in which case you should spend ten minutes or so each day playing with your cat), a triggering of the hunting instinct by a sudden movement the owner hasn't noticed, or perhaps a reaction to a change in the weather, barometric pressure, or whatever. Cats' bodies are so sensitively attuned to changes in their surroundings that they are aware of much that we will never notice!

Training

What cats can do

Why they will do it

What cats can be trained

Rules of behaviour

- owner's responsibility
- 'senior cat'
- where to start training
- cats on beds
- settling down a new kitten
- ground rules
- just say 'no'
- don't use punishment
- training aids
- claw-sharpening/scratching
- the scavenging cat

Training for 'tricks'

- how to train your cat to come when called
- teaching to miaow on command

Training

Some people still believe the myth that cats cannot be trained. Yet cats are not only more intelligent than any other mammal (with the possible exception of man and apes) but they have better memories, capable of remembering something shown to them once up to sixteen hours later. Some animal experts believe that cats will, in years to come, take over from the monkeys who have been trained as 'home helps' for disabled people. In future, it is thought that cats living with the disabled will be able to perform tasks for them, such as fetching and carrying for their owners, going to the shops for them and operating some types of machinery to assist them in their day-to-day life.

Cats have already been trained to perform in television commercials and in circuses – and even to perform magical tricks.

Trainers of cats say that a cat must see some advantage in carrying out an action he has been taught. If he can see no advantage, he will not perform. Being angry with a cat which refuses to carry out some taught action will have no effect at all; he will only 'work' if he receives some kind of pleasure from doing so. Consequently, trainers train their cats by rewarding them for their actions by giving them a type of food or treat which they are never given at any time other than during training periods.

Trainers agree that teaching should begin as young as possible, as young kittens' minds are more open to suggestion, although it is quite possible to train an older cat as long as you are patient and determined. They also agree that trying to train an unneutered or unspayed cat is next to impossible.

Given that cats are capable of learning to juggle or 'magically' pick the correct card out of a pack, it should not be difficult to train them in the simple rules of the household – the rules of behaviour which will make life for a cat sharing a home with a human owner more comfortable and easier for both.

The main obstacle to this sort of domestic harmony is usually the cat owner! Many will simply refuse to try to train their cats, believing it to be impossible. Others realise that training is perfectly possible but will allow their cats to do anything they wish because they fear their cats will not love them if they become stricter with them.

In fact, the opposite is usually the case. Because cats are not the solitary creatures of popular belief, living, in the wild, in social groups with a subtle gradation of dominance, there will always be a dominant 'senior cat' in any group – in your household, that senior cat should be *you*. By assuming this role, you are giving your cat a sense of security and identity. Give up this role and your cat will consider himself the senior. Once he has assumed this role, he will look at you in bewilderment if you belatedly try to assert your authority.

So, as with anything else, the secret of training your cat to behave in an acceptable manner is . . . start as you mean to go on.

When a new kitten or cat first comes to live with you, you will, of course, want to give him a few days in which to settle down before asserting that authority. During those first few days, allow him to do pretty well what he likes – with just a few exceptions.

Obviously, you will not want to let him do anything which might hurt him, so do ensure that your home is safe for him. Save wear and tear on your home – and your nerves – by blocking off open fireplaces (he can be told when he's settled down that they are out of bounds for exploration) and removing precious ornaments or valuable furniture.

But one rule you should institute from the very beginning. If you don't want your cat to spend every night for the rest of his life on your bed, then keep him out of your bedroom from the beginning.

Cat owners can be divided into two types – those who allow their cats on their beds and those who don't! A surprisingly large number of owners *do* allow their cats to sleep with them. In many cases, the cat is warmth and company (although this is sometimes resented by a spouse). But it appears that many owners see their cats sleeping on their beds as a symbol of safety and security. After all, at the first sign of danger, a cat would run away. Therefore to have a peacefully sleeping cat on your bed indicates that all is well.

Those owners who don't allow their cats to share their beds cite several reasons – hygiene being the most important, with a peaceful, undisturbed night's rest following closely behind.

Whatever you choose to do, it is vital to make that choice before the new cat or kitten comes to live with you – and to stick to your decision. If that decision is that your cat should sleep elsewhere, then the cat should become used to that from the very beginning.

Cats should not be allowed out at night (far less 'put out') as they are more liable to be lost, stolen or, at the very least, spend a cold, uncomfortable night out of doors. So find a safe, secure room for your new cat and provide him with a cosy bed, a litter tray and a bowl of water (you may wish to give him his evening feed in that room too). Cats *don't* mind being kept in at night; but should become used to it from kittenhood.

His bed can be as simple as a cardboard box with an old blanket in it. This will keep him warm and out of draughts. For the first few nights, especially if he is a kitten and is away from the warmth and comfort of his mother and littermates for the first time, you could place a warm hot-water bottle under the blanket. You could even place a ticking clock in the bed which will remind him of his mother's heartbeat.

Make sure he has everything he requires – then go away and leave him. As long as you are sure that the room is safe for him, with nothing to fall on him, cut him or choke him, *ignore* him if he cries. It may seem hard-hearted, but most kittens will not cry for more than one or two nights. Remember, if you give in, you could have that same cat sleeping on your bed for the next twenty years!

For the first few days, allow your cat or kitten to settle down without exerting much discipline. Give him time to come to trust you so that when you do eventually say 'no' to something, he will realise that you mean it. Also for the first few days, feed him whatever food he has been used to. However, if the food he is used to is inconvenient for you for some reason, or is simply not nourishing enough, begin to change his food to what you want him to eat – but gradually.

A sudden change of diet can lead to refusal or, worse, an upset tummy. Don't necessarily believe the cat's previous owner if you are told something like, 'He'll only eat salmon-flavoured Cattobounce', or 'He only likes bread soaked in gravy – he won't eat meat.' Very often these instructions mean that this is the only diet the previous owner has bothered to give the cat or kitten.

Of course, cats do have their own preferences, some of them deeply entrenched, but for a cat to enjoy complete good health his diet *must* be varied. Do remember that you are doing this for his own good when your pet turns pitiful, pleading eyes upon you because he doesn't immediately like the look of what is in his bowl.

Introduce the diet you wish your cat to eat at an early stage, having assured yourself that it is a nourishing and healthy one (see the section on food and drink). At first, you can mix a little of the new food in with the food that the cat is used to, increasing the amount of new food fed each day until your cat is completely weaned onto his new diet.

After the first few days, it is also time to start acquainting your cat with the ground rules of the household.

If he does something you don't want him to continue doing, possibly for the next twenty years, tell him so. All you need do is say – loudly and firmly – 'no' every time he steps out of line, at the same time gently removing him from whatever he is doing. Most cats are very smart and will learn very quickly what they can and cannot do. It is important to say 'no' as if you really mean it, and to discourage the behaviour each time your cat commits it. If he learns he will be told off sometimes but not every time, he will simply keep trying the behaviour on the grounds that occasionally he will get away with it.

Don't smack a cat for doing something he shouldn't; this is more likely to make him nervous than to make him behave. Some owners will wave a newspaper over a misbehaving cat's head; perhaps smacking it on the wall or against furniture at the end of its arc. This should be used as a last resort as, although the newspaper may not touch the cat, it will still upset a cat which has a tendency to be nervous.

If the loud, firm 'no' doesn't stop the behaviour within a couple of weeks, you could try the ultimate deterrent − a water pistol! If you keep one handy, loaded with clean water, and your cat does something you have told him not to do, a short, quick burst from the pistol will literally leave him wondering what has hit him. Most cats don't like getting wet so, if they get wet every time they perform a particular action, they will stop that action. And by using the pistol, they won't connect the 'punishment' with their owners.

Another, less drastic, way of persuading your cat to stop doing something he shouldn't − without getting him wet − is to spray him with compressed air such as that sold in cans for cleaning cameras. It sounds and feels like a hissing cat to another cat but, again, he won't realise you are responsible for it.

Do remember when training your cat that there are some instincts which you simply can't train your cat out of, for example, claw-sharpening. This is necessary for a cat for several reasons: it sharpens his claws (his weapons), it exercises him by allowing him to stretch his body and exercise the muscles which make his claws retract, and it also helps him mark his territory. If you don't provide your cat with something to scratch, he'll scratch your furniture.

Cats which have ready access to a suburban or rural environment will have trees on which to sharpen their claws but even they should have something indoors which they are allowed to scratch. An indoor cat, or one which lives in an urban environment, *must* have a scratching post of some sort.

Even a cat which has a scratching post always available will be tempted by furniture which is covered in rough, textured fabric and will often have a quick scratch as he passes by. If you have the sort of furniture which cats love, you'll have to bear with the occasional scratch on it, which should, of course, immediately meet with an oral reprimand.

Wallpaper can sometimes cause a problem too; for example hessian wallpaper cannot really be recommended for a cat-owning household – it's too attractive to climb!

In the early days, too, it is important that your cat comes to learn that *your* food is out of bounds to him. Don't give cats titbits outside of mealtimes, except as occasional rewards, as this will lead to a fat, greedy cat. It's best not to allow any cats in a room where any humans are eating. This stops begging, stealing and other similar behaviour and the cats don't mind – they quickly accept that certain rooms are out of bounds to them at certain times of day.

When cooking, shut the cats out of the kitchen, on grounds of safety and hygiene. Many cats have been badly burned by jumping on to cooker rings while they are still hot, so keep them out of the kitchen while cooking and for at least half an hour afterwards.

Cats can be scavengers and if you leave leftovers in the kitchen with the door open, don't be surprised if they're gone when you come back. You shouldn't try to train your cat out of this behaviour, it's too deeply ingrained and it would be actual cruelty to leave tasty titbits in front of your cat and not expect him to eat them. Train yourself instead that cats should be kept out of the kitchen until everything is cleared away.

Many owners feed their cats in the kitchen as this is probably the best place if the floor is vinyl or something similar. If the cat trails his food on the floor or spills his water, it's easily mopped up. But again, from the beginning, train your cat not to jump on to the work surfaces while you're preparing his food – or indeed at any time. Simply lift him down each time he does it, saying 'no' very firmly. Most smart cats soon realise that they will be fed quicker if you don't keep having to stop what you're doing to put them back on the floor. If they really won't stop jumping up, you'll have to lock them outside until each meal is ready.

Most cat training involves stopping a cat from doing something you don't want him to do. Few people wish actually to train their cats to perform actions, believing that this is degrading to an independent creature.

However, it is occasionally useful to be able to get a cat to do what you want him to do, if only to make it easier to take his photograph or fasten a collar round his neck.

Here you will find that giving a reward when the cat behaves as planned is invaluable. You will have to stock up with your cat's favourite treats. Mostly, these come in tubes or in cardboard tubs, all of which make a satisfying – and immediately identifiable – rattling sound when shaken.

Probably the first thing to teach your cat is the rattling sound of the treat packet. As soon as you have his attention and he comes to investigate, reward him with one of the contents. Usually, you don't have to repeat this for the cat to get the hang of it – virtually every cat will be able to identify the sound for evermore and come running!

This does have its uses if you need your cat for some purpose, for example, to give him a worming tablet, and he doesn't want to come. Do make sure he gets his treat *every time* he comes running in response to the noise. And, if you wanted him for something not particularly pleasant, such as the worming tablet, give him another treat afterwards to help him forget.

I also find it useful to train my cats to jump on to any piece of furniture I want them to. The signal for this is the rattle of the treat box followed by patting whichever place I want them to stand. So if I want to check my cats' eyes or look at a sore paw, I rattle the treat box and then pat a table, on to which they will happily jump, receiving their reward before I begin to check them over and another one afterwards.

Teaching a cat to miaow on command could be very useful in an emergency. If your cat was locked in a shed or trapped by a house fire, miaowing on command, no matter how frightened he is, could lead rescuers to him and perhaps save his life.

Choose a word to which you want your cat to respond – this could be 'speak' or 'hello'. You will then have to watch your cat and, when you think he is about to miaow, speak the word you have chosen. As soon as your cat miaows, reward him with a treat and tell him how good he is. Do this as often as is necessary to create a link in your cat's mind with the single word and his action and reward. When he truly miaows on demand, remember to continue rewarding him. It is very important, however, that he receives a reward *only* when he miaows in response to your signal. Otherwise, he will miaow all day long hoping for a treat.

There is practically no limit to the number of things a cat can be trained to do, as long as you treat him with great patience, going through steps in any complicated manoeuvres one at a time, and rewarding him every time he behaves as you wish.

Reasons for soiling:

- illness
- age
- litter
- dirty tray
- tummy upset
- tray
- lack of privacy
- confusion
- repellants
- stress
- change in routine
- nervousness
- sexual arousal
- allergic reaction
- territorial marking

How to clean soiled area

Soiling

This is the most common problem that cat owners have with their cats. By soiling, I mean cats squatting down to urinate (wet) or defaecate (pass a motion) indoors on furniture or flooring instead of in their litter tray or outside.

This behaviour can have a very serious cause – or a remarkably trivial one. If your cat behaves in this way, you will have to work through the list of causes, one by one, until all but one are eliminated.

The first thing to realise is that a cat *never* soils its home out of naughtiness or laziness. A cat behaving like this is telling its owner, in the only way it can, that something is wrong.

It is not only pointless to chastise a cat for soiling, it is cruel. You will never persuade a cat to stop soiling by smacking it or, even worse, rubbing its nose in the mess. In fact, this will reinforce the behaviour instead of stopping it.

Cats are extremely clean, fastidious creatures, so soiling on floors or furniture must immediately be recognised by an owner as a cry for help.

If a cat of previously clean habits starts relieving itself where it shouldn't, it is *vital* that the owner takes it immediately for a *veterinary check-up*. Cats are prone to a number of urinary conditions, including cystitis, and to problems such as impacted anal glands. But they are stoic creatures and will often not display symptoms until the condition becomes serious. So, as soon as a change in behaviour such as soiling is noticed, immediate veterinary advice must be sought. *Delay could cost the life of your cat*, while suitable treatment can often bring a swift cure.

When soiling has a veterinary cause, prevention is obviously better than cure.

Elderly cats, as might be expected, are somewhat prone to *kidney problems*, such as nephritis, so owners should be alert for the symptoms. These can include a tendency to drink more water than usual, and possibly lack of appetite, dull coat and pot belly. A vet will treat the condition with antibiotics and a change of diet.

Cystitis (inflammation of the bladder) is quite common in cats of all ages and may start off as a mild infection. When symptoms occur, these could include frequent urination of small amounts and possibly straining, with perhaps a cry of pain, even spots of blood in the urine, as well as soiling outside the litter tray. A vet will treat cystitis with antibiotics and possibly vitamin C, which will keep the urine acidic and less likely to harbour infection.

Feline Urological Syndrome (FUS) is an obstruction of the urethra by sand-like crystals. It has been estimated that around 10 per cent of cats suffer from FUS at some time in their lives. However, as only the most observant owner will notice the symptoms in a mild case of FUS, I believe that the incidence might be higher than 10 per cent. Symptoms can include straining to pass water or perhaps spots of blood in the urine. FUS can be treated in a number of ways by a vet, including manipulation or operation. Tablets may be given to acidify the urine to help disperse the obstruction and to stop new obstructions forming. There are also 'prescription diets' – available from your vet – which will also acidify the urine and prevent new crystals forming. It may be recommended (after treatment to remove the blockage) that salt is added to the diet to encourage the cat to drink more and help to 'flush through' impurities more frequently.

Impacted anal glands may also lead to soiling. When the glands on either side of the anus become impacted and lose their usual lubricating effect, the cat has difficulty in defaecating. The cat will frequently lick its rear end and a vet will treat the impaction by expressing the anal glands.

There could also be other veterinary reasons for soiling, so it really is imperative that a cat is first taken to the vet for a thorough check-up.

Old age is commonly a time when soiling inside the house may begin. This is not due to cussedness, as so many owners of elderly cats seem to believe – the cat simply can't help it. As he becomes older, a cat may begin to lose control of his bodily functions – excretion being one of them.

Don't tell him off – that will do no good and simply make his old age a miserable time – and don't keep putting him out because he's now 'dirty'. Try to remember all the good times he has given you, the fun and companionship you have shared, and attempt to deal with the problem philosophically.

Provide extra litter trays, both upstairs and down, and, if there is one particular spot he has chosen to relieve himself, place a tray there if at all possible.

Take him to your vet for regular check-ups, as his

other bodily functions may also be failing. But above all, treat him with all the patience you can muster.

There are many reasons for soiling in the younger cat.

Only about 30 per cent of cats in the United Kingdom are provided with a litter tray, but this is an essential piece of equipment. Although cats are very continent creatures, it is unrealistic to expect a cat always to wait for you to let him outside before he can relieve himself. This is particularly true of the older cat, whose bladder may not be working as efficiently as it once was.

Owners should take on the responsibility of checking their cats' waste products, as these can often give the first indication of urological problems, worms, diarrhoea, etc. So a litter tray will often provide the first clues to an illness, when the owner notices spots of blood in the waste products, or sees the cat straining to pass liquid or solid waste.

If you have provided a litter tray, but your cat is refusing to use it, this may be for a number of reasons.

Has the *type of litter* been changed recently? There are a number of different types of litter on the market and some cats have preferences for particular types. The most common types are natural mineral clays, dried and crushed. These include Fullers' Earth, also known as Grey, which is grey in colour and heavy in weight and is produced in Britain. Sepiolite is a white mineral litter which is lighter in weight and comes from Spain. Moeler clay is also used as a litter, is pinkish-brown in colour, and comes from Denmark. Then there are pelletised softwood litters, made of desiccated spruce and pine formed into pellets.

Trouble can arise when the type of litter is changed,

as the conservative cat may not like what he smells, or feels under his paws. So if you are changing the type of litter, place a little of the clean, old-type litter on top of the new, until your cat has adjusted. If he doesn't adjust, he is simply pointing out that he doesn't like the new type of litter. A few types of litter have deodorants added and some cats don't like these, so check the litter bag to find a litter without additives.

In the event that none of the mineral or softwood litters meet with approval, there are other substances which can be used instead. Peat makes a good cat litter and is easily disposed of on the compost heap, although it may lead to black pawprints through the house. Newspaper can be used if necessary, although this will have no natural deodorant properties and may also lead to black pawprints. Shredded kitchen paper is another alternative, and this is useful when the vet requires a urine sample; the shredded paper is simply squeezed to wring the urine into a sample bottle. Some people have even utilised smooth, clean pebbles as a litter medium and these will work adequately if you and your cat have no objection, although you will have to wash the pebbles regularly.

The cat's natural predilection for *cleanliness* might also lead to soiling outside the tray. Cats are just as fastidious as human beings and many humans wouldn't dream of using a dirty toilet. Yet some owners expect their cats to use an already soiled tray. Many simply won't. If the tray is dirty, they'll find a nice clean piece of carpet and relieve themselves there instead and, if they do, you have no one to blame but yourself.

It is often recommended that litter trays should be cleaned 'daily'. In fact, there should always be at least one spotlessly clean tray in your home at all times. For owners with one tray, this means cleaning it as soon as it is used (just like flushing the toilet after every use).

Some cats are more fastidious than others but few, understandably, will enjoy using a dirty tray, so you may suffer from a 'once-only' soiling if you haven't been quick enough to suit your cat.

Soiling may also occur if your cat has a tummy upset and just can't get to his tray in time. This may be because of something he has eaten, or even because he has been given too much milk to drink (milk can cause diarrhoea in many cats). Watch his diet and change anything you feel might be upsetting him and this should not recur.

If you have more than one cat, you should have more than one litter tray. Some cats will refuse to use a tray another cat has used, even if it is clean at the time. (With their excellent sense of smell, they can tell that another cat has used it and may look on that particular tray as the other cat's territory, since urine and, to a lesser extent, faeces, are used to mark territory.)

The tray's situation is also important to cats, who may refuse to use a badly sited tray. It should be placed in a quiet, traffic-free spot where the cat can use it undisturbed. Some cats will not want to use a tray which is sited in the same room as their feeding bowls, or even their bed. In the wild, a cat will defaecate well away from its nest in order not to attract larger predators by the smell.

Some cats will insist on choosing the exact spot where their litter tray should be situated, and the best thing is to allow them to do so if it is not too inconvenient to other members of the household. One of my cats insisted on urinating in one particular corner of the kitchen. A veterinary check-up showed there was nothing wrong but she continued to wet in that corner. After I placed a litter tray there she used it happily and never soiled outside it again.

Sometimes, a cat will soil somewhere you simply can't place a tray, for example, in the middle of the living-room carpet. Either you will have to place a piece of furniture there, or else place a litter tray there temporarily, moving it a few inches every day until it is situated in a convenient place. Or you can try placing your cat's feeding bowl on the area – at least temporarily – as a cat won't soil his feeding area.

Some cats like *privacy* when using their trays and will prefer their tray to be situated in an out-of-the-way place, such as the cupboard under the stairs. If you don't have a convenient hidey-hole which you can use, you can place the tray inside a large, up-ended cardboard box with a hole cut in the side to allow your cat access. There are also covered litter trays on the market which provide purpose-made privacy.

A cat may start soiling on a carpet, say, simply due to *confusion*. A cat will always 'cover up' what he has done in the litter tray and sometimes his paws will be *outside* the tray making the scraping motions of covering up. If the cat feels carpet under his paws, this can become confused in his mind with his litter medium, and he will begin to use the carpet as a toilet. You can overcome this by placing a large, cut-down cardboard box or a sheet of plastic under the litter tray.

Cats will usually not soil on plastic, so if your cat is soiling in one particular place, for example, your bed, cover it with a sheet of plastic when not using it and the behaviour should stop.

A cat may soil in the wrong place if his *tray has been moved*. If you move its position and your cat continues to go where the tray used to be, you will either have to give him time to get used to the new arrangements, or move the tray back. Another of my cats will leave a neat little pile on the bathroom carpet in the exact spot his tray used to be – more than a year after the tray was moved!

Trays should not only be cleaned regularly, they should be disinfected too. The *disinfectant* you use on the tray might be repelling your cat, leading to soiling. Many disinfectants and antiseptics which are harmless to humans and other pets can be toxic to cats. Read the labels on disinfectant bottles. Don't use anything containing chlorinated phenols or coal tar products and it is best to steer clear of 'pine' disinfectants too (although litters made from natural pine woods are perfectly safe). Don't use Jeyes Fluid or Dettol. The best disinfectant to use is a diluted solution of sodium hypochlorite, which is readily available in the form of Domestos and similar products.

You can ensure that the tray is not smelling of anything which your cat will find offensive (many are repelled by the smell of urine) by rinsing it in a solution of baking soda in water. This should help to neutralise smells.

Soiling may also be caused by *stress or a sudden change in routine*. Cats are creatures of habit and a change may not agree with them. So if a previously non-working owner starts working and leaves their cat alone, or a member of the family leaves home, or there is an addition to the family (adult, baby or animal) or a change of residence, your cat may start soiling somewhere he shouldn't. This is not unlike a young child beginning to wet the bed because of an alarming change in its life. As with the child, patience and comforting are the answer here. Don't, of course, shout at your cat as this will upset him further and reinforce the behaviour. Spend a few minutes each day with your cat quietly stroking him and making him feel the most important creature around once again.

Nervousness could also be a cause of soiling. A very shy, nervous cat may simply not be able to help himself. Again, patience and understanding will be rewarded with a cat which is more confident and less likely to soil.

Female cats *on heat* will urinate indiscriminately so, if your female isn't a pedigree from which you intend to breed, have her spayed. Some spayed (or neutered) cats will continue to behave in this way if the spaying (or neutering) hasn't been completely carried out and part of the reproductive organs remain. Only your vet will be able to tell if this is the case.

There can be strange reasons for soiling too. A few cats will continually soil their own beds, if these are in the form of beanbags. At first glance, this seems completely rogue behaviour as most animals, and especially cats, will never soil their own beds. Confusion is the answer here too, I believe. Beanbag bed filling sounds exactly like cat litter if scratched, and it must have a feel under the paws very similar to litter. The cat who soils his beanbag doesn't realise he is doing so – he believes he's on his litter tray. After all, at least two of his senses are saying so.

Another unusual reason for soiling may be the *food you feed your cat*. Most manufactured cat food contains colourants and preservatives and, although the majority of cats suffer no ill-effects from manufactured food, there is the occasional one which seems to react badly to it. Just as food additives have been shown to provoke unacceptable behaviour in humans, it seems that some *cats* may be similarly affected. In these cases, a simple change of diet – to a different brand of food or to food prepared by the owner – will be enough to stop the problem.

Soiling may also be caused by some substance in the environment which a cat may react to badly – for example, it may be a reaction to a carpet-cleaning solution, new paint, or some other equally 'innocent' substance which the owner may not suspect.

Some people believe that a cat which is soiling should be placed in a cattery for a few days as this often stops the problem. This is certainly worth a try, but bear in mind that it may be either the *change of diet* or the *change of location* which is responsible.

Soiling may be a *territorial* behaviour too. Although most cats who wish to mark their territory will do so by spraying, some cats don't quite get the hang of spraying and may urinate or defaecate instead. (More information about the territorial instinct can be found in the next section.)

Whatever the cause of soiling, it is vital that the soiled area is *thoroughly cleaned*. If any smell remains, the cat will be triggered by the scent to use the area again. So wash the area with clean water, then with a weak solution of Domestos and water to kill germs. (Remember that Domestos is a bleach and could bleach your flooring or furniture. Try it first on a small, out-of-the-way area.)

Then, to completely eradicate smells, sprinkle the area with dry baking soda. This will absorb some water from the cleaning, so allow it to dry for a day or two, then vacuum up.

A vinegar rinse, after washing the soiled area, is often an effective deterrent, or a sheet of plastic or other obstacle temporarily placed on the area will discourage further use. Other repellants can also be tried, such as oil of peppermint, citrus oils or even pepper.

Common Problems   SPRAYING

How to recognise spraying

Reasons for spraying:

- cat is still entire

  territory

  age for neutering

- neutering incompletely carried out
- overcrowding/territorial marking

  hierarchy

  separate areas

- stress

  don't shout!

- presence nearby of another cat
- in response to a female

How to deal with spraying

- living with it
- rehoming a cat
- making a run for the spraying cat
- converting a room for the spraying cat

Cleaning the area

Spraying

This is the second most common problem which owners have with their cats. A cat will spray by backing up to an object, raising his or her tail high and spraying backwards a jet of strong-smelling urine. While spraying, the tip of the tail may wiggle back and forth. Contrary to popular belief, both males and females will spray, although females are less likely to spray than males.

This is territorial behaviour: a spraying cat will patrol the boundaries of its territory regularly, spraying (marking) the same selected areas each time.

It is thought to be a 'warning' to other cats and it will certainly be used to overlay the smells of other cats. It can equally well be used to make strange items smell familiar to the spraying cat.

At a recent visit to a cat's home, for example, I left my handbag sitting on the floor. Of course, the resident cats came over to sniff it and found it smelled 'foreign'. To make it smell familiar, a number of them sprayed it and I came back to find it dripping wet!

Problems arise when some cats choose to spray in the house. Some owners simply live with this, as we shall see later, but of course the majority of owners would like to know how to stop it!

A neutered or spayed cat is less likely to spray than an entire (unneutered) cat. If you bear in mind that spraying is mainly territorial in origin, it is interesting to note that neutered and spayed cats have *smaller* territories than entire cats. An entire male will have a territory which will include your home and may be as big as 150 acres, and he will lay claim not only to the food sources within that area, but also to any unspayed females who will have him! A neutered male's territory will be much smaller than this but will still be six or seven times that of a spayed female.

So not only will a neutered cat be content with a smaller territory, but the effect of the operation will be (in the male) to lower the levels of testosterone in the cat's body. It is this hormone which can trigger spraying behaviour.

It is to be hoped that every non-pedigree cat and the majority of pedigree cats are spayed and neutered anyway to prevent unwanted kittens, but if not, spraying by the cat should be enough to convince any owner of its necessity.

Females can be safely spayed at any time from four months onwards (take your vet's advice on the best time to spay) but they are less likely to spray indoors anyway. The problem usually arises with male cats and this is where an owner's powers of observation and knowledge of their cat becomes important. It used to be recommended that a male cat be neutered at any time from four months onwards, but now many vets are recommending that tomcats be neutered somewhat later. This not only allows the male hormone, testosterone, to be manufactured in the cat's body, making him bigger, stronger and with a well-developed bone structure, but it minimises the risk of the development of urological problems in later life.

However, an unneutered cat *is* more likely to spray, starting usually when he is a 'teenager', around eight months of age or so.

Owners should watch their male cats carefully and help their vets to judge the best time to have their tomcats neutered. As a rough guide, eight months seems a good time, but the maturity rate of cats varies tremendously as they are so individual. If you find your tomcat is becoming more assertive, and perhaps more interested in what is happening out of doors than in your own home, and it coincides with an age of roughly more than eight months, it is perhaps time to consult your vet.

Tomcats recover very quickly from neutering, which is a very minor operation, and, within half an hour of their return from the vet's will be jumping around and demanding the breakfast they didn't have! (Cats should not be fed on the day of any operation. It is also important to remember that they should not have been de-fleaed near to the time of any operation as this can prove toxic in combination with the anaesthetic.)

It's important to try to stop a cat spraying before the behaviour is established, as afterwards it will be more difficult to curtail. With luck, spaying and neutering will prove sufficient to prevent the behaviour.

Spraying behaviour can sometimes occur in a neutered cat because some *testicular tissue remains* after the operation. This will continue to produce testosterone and the cat will continue to behave like an unneutered tomcat. A veterinary visit will be necessary to monitor the levels of testosterone and if any tissue remains, this can be removed surgically or hormone levels can be adjusted by a course of drugs.

A problem sometimes arises when an owner has *more than one cat*. This is seen as a territorial threat within the home by some cats who, rather pathetically, resort to marking items of furniture or boundaries within their own homes.

An example of this can be seen if you visit any shelter where a number of cats are being kept before rehoming. Despite scrupulous hygiene, there is usually a strong smell of urine pervading the place. This is the result of large numbers of cats being kept in one area, and a number of the cats deciding to mark out territories for themselves.

The probability of cats in a multi-cat household behaving in this manner can be minimised in a number of ways. First of all, if you are choosing a new cat or kitten and you own at least one already, try, if you possibly can, to choose a cat whose personality will fit in with that of your existing cat. If your cat is assertive and likes to be boss, don't bring home another assertive cat. Try to choose one who will take his place second or third in the hierarchy – because there most definitely *is* a hierarchy among cats.

Difficulties arise when several cats are equally assertive and each one wants to be 'chief cat'. It will help to calm things down a little if *you* ensure that *you* are 'chief cat' in your household and arbitrate in disputes before they have gone too far.

Of more immediate use will be ensuring that each cat has his 'own space'. Each cat should have a bed of his own and preferably his own litter tray. Feeding bowls should be kept separate and it is not a good idea to feed a number of cats from one bowl. Apart from increasing the risk of spreading illness or disease, there is always one cat who won't get enough to eat. A new cat in the household should have a bowl separated by a distance of some feet from any other cat's bowl. In a feral group, each cat will stay at a distance of several feet from other cats, the distance decreasing as the cats get to know and trust each other.

There is a 'danger' number of cats in a multi-cat household. Many spraying problems arise when there are *four or more cats* sharing one home. Of course, this will vary, not only due to the temperament of the cats but the size of the home. Someone with a large house where their cats have a large number of places to play, hide and generally 'get away from it all', will be less likely to have a problem than someone with the same number of cats living in a one-bedroom flat.

Giving the cats separate areas of their 'own' should help to discourage spraying. If one cat is rather assertive and another, shyer cat is intimidated by this behaviour, give the shy cat a 'bedroom' of his own and that will immediately cut down on many problems.

I have four cats: one very assertive female, one fairly assertive but good-natured male, one shy female and a male who will go along with anything for a quiet life. The good-natured one has his own dining-room as he eats slowly and will happily stand by and watch while someone else eats his food for him. The assertive female also eats alone, while the other two eat together. At bedtime, the two males share a room and happily cuddle up together. The shy female has her own 'bedroom' with her own bed about eight feet in the air, on top of a stable piece of furniture (it gives a shy cat more confidence to sleep high up, as they would in the wild). The assertive female has the whole of the rest of the house at night-time! This may sound complicated, but a system such as this can keep a variety of personalities living contentedly together with fewer confrontations or arguments. However, at holiday times, it does drive cat-sitters crazy!

Stress of any kind can lead to spraying; in fact, it is the second most common reason (after hormonal reasons) for spraying behaviour. Moving house, new additions to the family, someone leaving home – even feeding in a new area – in fact, change of any kind, can lead to stress-related spraying. If the behaviour can be traced to a change in routine and that change can be reversed, this should be enough to stop the behaviour. If the change in the household is irreversible, patience is the only answer. The behaviour should modify or stop once your cat has become used to the change and will stop quicker if you can restrain yourself from remonstrating with him about the spraying! If you shout at him, this will simply add to the level of stress he is suffering and make the behaviour continue for a longer time than it otherwise would.

There can be reasons for spraying which are not immediately apparent to the owner. It could occur in response to a *threat* from another cat of which the owner might not be aware. Other people's cats will almost certainly spray in your garden, if you have one. You may not be aware of this, except when the smell is particularly bad, but your cat will. He will be desperate to overlay the smell of the other cat with his own. This can often lead to a cat spraying near or around the doorways or cat-flap to his own home. The other cat may have sprayed just outside the door, but your cat, unable to get out, may spray just inside.

He may be simply overlaying the smell of another cat when you have no idea that another cat is involved. For example, one house cat always behaved impeccably until he started spraying on the coal in the coal scuttle. This behaviour mystified his owners until it was pointed out to them that their coal shed was left open at all times. A neighbour's cat was spraying on the coal in the shed and when it was brought into the house in the scuttle, the house cat could smell the neighbour's cat although his owners could not. He was simply overlaying his own scent on that of the invader of his territory. Once his owners knew what the problem was, they kept the coal shed closed and the behaviour ceased.

So, if one particular item is frequently sprayed, ask yourself if another cat could be involved, and, if so, take steps to prevent the *other* cat from spraying.

If you've never seen your cat spray, yet you come home to that awful smell which indicates that spraying has taken place, don't immediately jump to the conclusion that your cat is responsible. If you have an open cat-flap or window, you will find that many neighbourhood cats will take advantage of them in your absence, coming into your home and spraying over anything, as a sort of defiant 'Kilroy was here' gesture. This can only be stopped by keeping flaps and windows locked, or fitting an expensive electronic or magnetic cat-flap which will only open selectively, in response to a gadget on a collar worn by your cat.

A male cat may also spray *in response to a female* in season. He will be able to smell the female from some distance away and, if a female in season comes into his territory and he cannot get to her, he will respond by spraying. This, of course, will be the natural response of an entire male; neutered males don't usually respond in this way, but can do so, on rare occasions.

Not everyone who lives with a spraying cat takes steps to prevent spraying. The owners of entire male pedigree cats, for example, consider spraying an occupational hazard. If their cats are stud cats which live as part of the family, they will simply go around their house each day with a damp cloth, wiping up anything they find.

For the average owner, this is not always the best answer. If your spraying problem is related to having a large number of cats, you should consider rehoming one or two of them with a sympathetic owner. Very often, this will be sufficient to stop the behaviour. If you are unable or unwilling to rehome the cat or cats, and giving them their 'own' territory within the home has not helped, you should then consider giving them space of their own outside the home.

If you have a garden, a cat run with or without a cat chalet could prove useful for giving a cat or cats territory of their own. Cat runs can be purchased ready-made from specialist manufacturers who advertise in cat magazines. The runs are fairly inexpensive, made of a framework of wood, completely covered in wire mesh, and the smallest ones measure 6ft × 6ft high × 3ft wide. Cat chalets can also be purchased from the same manufacturers and these will attach to the end of the run with access for the cat from one to the other. A chalet and run could also quite easily be built by a DIY owner.

If your cat is to spend just part of each day there, you will need little additional equipment other than a few toys, his feeding bowls and litter tray. If, however, this is to become your cat's semi-permanent home, you will need to run power to the chalet for heating in cold weather and lighting so that you can see what you're doing when you visit your cat at night. For you *should* visit your cat frequently; sit in the run, talk to him and play with him. He has not been put in some sort of jail for bad behaviour; he has simply been separated from the other cats to give him space of his own and to keep him happier. But cats do suffer from loneliness – so do remember to visit him and spend as much time with him as you can.

Every so often, you can try putting him back in the house with your other cats. You may find that his spraying has stopped or modified.

Other people deal with the problem in other ways. One friend of mine has a number of cats, of which two spray. They are put out to play in the garden in the daytime and given a room of their own in the house at night. This room has been totally converted to their needs. Everything in it is washable – ceilings, walls, floors, furnishings. The floors are covered in vinyl, taped at the edges to the skirting boards so that no liquid can escape. After a night in their own room, the two cats are put into the garden while their owner thoroughly washes their room ready for the following night.

When a spraying problem is not as extensive as this, it is possible to try to deter a cat from re-using a specific area. Clean and disinfect the area thoroughly, then cover the area with a substance which cats don't like, for example, citrus oil, oil of peppermint or vinegar.

Deterrents do have a limited effectiveness, however. It is much better to try to understand the reasons for spraying and make some modification in your way of life which will lead to your cat deciding not to spray indoors any more.

If all else fails, it has even been known for some owners to place a baby's disposable nappy on their cat – with a hole cut out for their tail to go through!

What a cat looks for in a food

How to choose a food

Variety

Fresh food

Prey

Fish and liver

Changing diet

Multi-cat households

Cadging cats

Messy eaters

Cleaning bowls

Ill cats

Hot weather

Dog food and vegetarian diets

Drinks

Food and drink

Most problems with cats tend to relate to one end or the other! The greatest problem most owners have with their cat's eating habits is when their cat will only eat one food.

It is interesting to note that what a cat will look for in a food bears little relation to what an owner will look for in a food for their cat. Many owners will buy a brand of tinned food for their cat to try once; then, if the food has an unpleasant (to the owners) smell, they will not buy that brand again. However, cats enjoy strong-smelling food and they will probably prefer the brand their owner dislikes.

Apart from smell, the temperature the food is served is very important to a cat. In the wild, they would catch and kill small prey animals, which they would immediately eat. Obviously, then, they will prefer food which is served at blood temperature, rather than food which is straight out of the fridge.

Palatability is also important. With convenience foods, manufacturers spend a lot of time and money finding foods which cats will find palatable, although most cats have individual preferences about particular flavours.

Nearly all convenience foods have colouring added – not for the benefit of the cat, who won't care if the food is brown or green, but to appeal to the owner.

It is important, therefore, to find a nutritious food *which appeals to your cat*. If *you* don't like its look, texture, colour or smell, put aside your reservations. After all, you don't have to eat it!

It is absolutely vital to feed a cat a varied diet from kittenhood, so that the kitten doesn't grow up to be a faddy, fussy cat. It is during kittenhood that you should do your homework and study the labels on cat foods. Also decide at that stage which fresh foods you will feed your cat and how frequently, and start the kitten off with the diet you intend to continue throughout its life.

Tinned food and packet food have their ingredients printed on the sides and this (along with price) should form the basis of your choice. Usually you get what you pay for in cat food; the most expensive foods will contain a larger proportion of the named meat, while the cheaper ones may simply be 'flavoured' (containing little or none of the named meat). The cheaper ones will also contain cereal for bulk.

It doesn't always follow that a certain food is better than one which is *slightly* cheaper – and this is where the label-reading is important. If you can find a food which is high in protein (for growth), low in ash (high ash levels have been implicated in the incidence of Feline Urological Syndrome) and with little or no cereal added, *and your cat likes it*, it would probably be worth sticking to that brand for any convenience food meals you feed.

Cats have very specific nutrient requirements and you should read the labels to see if they are designated a 'complete' food. If so, theoretically, the cat could be fed nothing but this food and lead a healthy life, although I prefer to feed a variety of foods, both fresh and convenience, as not all the nutritional requirements of cats are yet fully understood. A 'complementary' food will not provide all the nutrients a cat requires and I prefer not to use these foods as I like to feed as near as possible a 'complete' food (fresh or convenience) at every meal.

Some owners believe that feeding a varied diet means changing from one brand of cat food to another. Whereas this might be worthwhile to an owner if cash is short and special offers can be taken advantage of in this way, it is not of particular benefit to the cat. It will certainly do him no harm as long as the various brands are of good quality, but fresh foods should also be given.

I find it best to stick with one brand which I believe to give the best combination of nutrition and price. Several times a week, I also feed fresh food: cooked fish (white *and* oily), liver (*no more* than once a week), raw mince as a treat (mince suitable for *human* consumption only), fat trimmings from my own meals (both cooked and raw) and *good quality* scraps (dark meat, giblets and skin from cooked fowl, scraps of hard cheese, etc.).

My cats will also catch and eat small rodents from the garden. Although I obviously don't encourage this (apart from anything else, the prey could be poisoned or be carrying a disease or cysts leading to worms), if my cats do catch and eat prey, at least they are eating their natural food and probably obtaining nutrients difficult to find anywhere else. (For example, cats will obtain folic acid and vitamin K, the anti-coagulating vitamin, from vegetation in the gut of their prey.)

It is also important not to allow a cat to insist on eating a fresh food to the exclusion of all else. Liver, for example, contains large amounts of vitamin A. Some cats become 'hooked' on liver, leading to a vitamin A overdose, which can cause bone deformities. Fish, also, which many cats are frequently fed, does not, on its own, constitute a balanced diet. Although fish is a good addition to a cat's diet, he should not be fed fish exclusively. (Fish should always be served lightly cooked. Raw fish contains an enzyme, thiaminase, which inactivates thiamin – also called vitamin B1 – a very necessary nutrient in a cat's diet.)

I have received letters from worried owners who say that their cats will only eat 'fish' flavour tinned food, and will this do them any harm? Although *fresh fish* will not contain all the nutrients a cat requires, a 'fish' tinned food will contain little or no fish, some being no more than fish-flavoured. So, as long as a particular tinned food is marked 'complete food', this shouldn't cause problems – although it is obviously better to feed convenience *and* fresh foods.

When *changing your cat's diet*, it is always necessary to change gradually. A sudden change from one food to another, if the cat will accept it, will most likely cause diarrhoea. If your cat resists the change, start adding the new food to his old food in small amounts and mix thoroughly. Increase the amount of the new food each day until he's eating the food you want him to eat. Resistance can be overcome by pouring an attractive liquid over the food. For the cat who is overfond of fish, a fish stock can be poured over a less-attractive food to encourage him to eat. Many cats will also enjoy a beef stock, stock made with yeast extract, or stock made with giblets. Always pour over the food and mix thoroughly (it doesn't matter if it looks sloppy) as cats are expert at lapping out attractive food and leaving less-attractive food behind!

If a cat is particularly finicky, he may disdain the contents of his bowl while pleading to be fed his favourite food. If you're attempting to alter his diet, you will simply have to try to ignore him. Leave the food in his bowl until it is eaten (unless it goes off first or is contaminated by flies) but if he hasn't weakened within two days, you will have to try a different food.

It sometimes helps with a finicky eater if his food bowl is removed ten minutes after it has been filled and not replaced until the next mealtime. Most cats will get the message that they have to eat *what's* there *while* it's there. Fussiness about food doesn't often occur in a multi-cat household as the finicky eater will know that anything he picks over will quickly be gobbled up by a less-finicky eater.

In multi-cat households, it is very important to ensure that the cats actually receive all of what is in their bowls. Some cats are like vacuum cleaners on full power, scooping up their own meals in seconds, before starting on someone else's. The slow eater in a multi-cat household can lose out. Cats should always be given their own feeding bowls and should either be watched while they are eating, or placed in different rooms if unmatched in speed of eating. Mother cats need to be watched particularly; they will often defer to kittens, allowing the kittens to eat their food, and sometimes not receiving enough nourishment themselves. This behaviour can continue even when the kittens are full-grown, so it is important to ensure that they are given peace to eat their food themselves.

Full-grown cats can be fed once or twice a day; I prefer to feed mine twice a day. If the last meal is late in the evening, it is an incentive for a cat which is outdoors to come in and spend the night indoors. A late feed will also encourage cats to settle down and sleep, allowing the owner an undisturbed night's rest.

Some cats receive more than two meals a day – unknown to their owners. These are the cats which have got cadging down to a fine art. They will eat their breakfasts at home, then go on a tour of their area, miaowing plaintively at kindly neighbours' doors until they are fed again.

Ask your neighbours not to feed your cat but, if you suspect someone is, and your cat is becoming too heavy, feed him food he's not particularly keen on – he'll only eat it if he is really hungry.

One person in each household should be responsible for feeding the cat – and only one. Otherwise, many cats are fed over and over again by each member of the family.

Many owners are annoyed by cats which pull food out of their bowls onto the floor. There can be several reasons for this. The bowl may be too narrow for the cat's whiskers; cats prefer bowls which are wide enough to allow room for their whiskers. Or the food may be in too-large chunks. Cats will pull large chunks of food onto the floor to 'chop it up' with their teeth. Or perhaps the bowl smells – either of old food, or of whatever you use to clean it. Bowls should be regularly sterilised with a weak solution of sodium hypochlorite and water. No other sort of disinfectant should be used on bowls, and some cats will even object to washing-up liquid.

On the subject of feeding bowls: some cats will have an allergic reaction to some plastic types and may develop a rash on their chins. In this case, ceramic or stainless steel bowls should be used.

It is possible that rashes around the area of the mouth could be caused by a food allergy; if your vet is unable to clear up the problem, you could try changing your cat's diet, one item at a time, to see if this has any effect. If you give your cat an egg regularly, that could be the culprit, or an allergy to milk or even beef might be the problem.

Ill cats are often a problem to feed. A cat needs to smell his food to be attracted to it and often a sick cat won't be able to smell his food. Try feeding strong-smelling food such as mashed-up sardines, or adding beef or yeast extract to his usual food.

If an ill (or old) cat has difficulty chewing his food, you could try liquidising it for him with a little stock, water or milk. He may find it easier to lap up than to chew. If he's unable even to lap, the liquidised food can be placed in a syringe (with the needle taken out and disposed of carefully). Release the liquid *gently* between the lips in the side of your cat's mouth. Obviously, don't push the syringe down his throat and do feed very small amounts at a time, so as not to choke your cat.

Many cats will go off their food during hot weather and will not eat. Many owners worry about this, spending time and effort trying to persuade their cats to eat and tempting them with tasty titbits. This really isn't necessary. If a cat refuses to eat because of hot weather, he is behaving quite naturally and really doesn't need food.

Eating will generate more heat in his body and he

will be spending as much time as possible keeping cool. If cats are fed twice a day, feed the first meal during the very early morning when it is still cool. Refrigerate any leftovers which should be fed (after being allowed to warm up to room temperature) in the cool of the evening, say, around ten o'clock or so. A cat is more likely to eat at this time of day. Do ensure that there is a plentiful supply of fresh water available at all times.

Some owners feed their cats dog food for economy as it is appreciably cheaper than cat food. A cat will not have a healthy life if fed solely or mainly on dog food, and will probably go blind. The cat has a much higher nutrient requirement than a dog, which accounts for the higher price of cat food. Manufactured cat food includes, for example, taurine, an amino acid vital to a cat for full health. Deprived of taurine, a nutritional deficiency will set in, leading to retinal atrophy and blindness. For the same reason, cats cannot survive on a vegetarian diet. (Dogs can manufacture taurine within their bodies.)

Some cats do like to eat foods which would not normally be associated with the feline species. For example, one of my cats adores asparagus and cauliflower and will ignore meat in favour of either. These sorts of foods, chopped up and fed in moderation, should do no harm, but do keep an eye on your cat's litter tray to spot any harmful after-effects! A little of what they fancy does them good, but sweet or starchy foods such as chocolate, cake or biscuits should not be fed as they will simply fill the cat up without providing any nourishment.

Drinks often bother many owners. They ask how they

can persuade their cats to drink milk. The answer is –
don't bother!

Once weaned, cats don't need milk as long as their
diet is otherwise good. It is a completely unnatural food
for any adult mammal and some cats simply don't like
it. Other cats are even allergic to milk (particularly the
foreign types such as Siamese or Rex) and it will give
them diarrhoea.

If concerned, you could give your cat a calcium
supplement, being careful not to overdose. Special
calcium supplements for cats are readily available from
vets or pet shops, or bone meal can be given – but do
ensure it is the edible type, *not* bonemeal used for the
garden, which is not sterilised.

It is very important that cats drink plenty of *liquid* but
water is probably the best liquid for them to drink, and
is especially important if dry food is given.

Many cats won't drink much water, so it's a good idea
to encourage them to do so in the following ways.

Some cats will ignore their water bowls and drink out
of their owner's baths or from a dripping tap. It may be
that these cats prefer not to drink near their food bowls.
In the wild, cats would not eat near their waterhole,
and they will drink from running streams.

The chlorinated smell of most water can be offputting
to cats. The smell will disperse if you bottle the water for
a few days before using it. Or you could try filling your
cat's bowl with cooled, boiled water. A small amount of
honey added to the water can also be enjoyed by some
cats. If you live in an area where the rainwater is
unpolluted, you can catch that and put it in your cat's
bowl.

If your cat simply *won't* drink water, make up a plain

stock from giblets or bones, strain, cool and serve.

And if you are away from home for a few days with your cat. take a supply of bottled water with you. Water varies in different parts of the country and your cat will be happier with what he is used to.

Common Problems
'AGGRESSION' TOWARDS AN OWNER

Reasons for aggression:
- previous ill-treatment
- illness or injury
- instinctive reaction

  how a cat fights
- restraint

  why a cat uses its teeth

  owner's responsibility

  laying down your guidelines
- ambush

  how to stop the ambushing cat

How to deal with other types of aggression
- anti-stress vitamins
- hormonal imbalance
- tranquillisers

'Aggression' towards an owner

What is often seen by an owner as 'aggressive' behaviour directed against them is usually instinctive behaviour which is misunderstood by the owner.

'My cat bites me,' is a fairly frequent complaint but this behaviour usually has nothing whatsoever to do with aggression.

Those few cats which are truly aggressive usually become so because of *previous ill-treatment* (although ill-treatment is more likely to result in a nervous cat than an aggressive one). In this case, enormous patience (possibly over a period of several years) and kindness are the only answers.

A cat may also begin to act in what its owner sees as an 'aggressive' manner because it has had *an injury, or because it is ill.* This is where knowing your cat thoroughly becomes so important; as soon as you notice a change in behaviour, a thorough check-up should be done to see if there is a physical cause for the change.

If your cat is healthy and normal, and bites you when you *tickle his tummy*, his behaviour is not directed against you. Cats will work out a hierarchy amongst themselves, often decided by fights or playfights. The cat's best fighting position is lying on his back, enabling him to use all four paws to fend off attack and leaving the mouth free to bite, if necessary. The cat which is fighting on top will scrabble away with his back legs on the abdomen of the cat underneath. This 'scrabbling' is duplicated by the owner's tummy tickling, which will trigger the fighting instinct and the owner may be bitten. This biting is often little more than a gentle pressure of the teeth on the owner's hand, often followed by an apologetic lick when the cat 'comes to his senses'.

It is very easy behaviour to stop – just don't tickle your cat's tummy.

Another type of biting, again usually no more than a gentle pressure of the teeth, is a *restraining* bite. Again, this type of bite is little more than a grasp by the teeth, without breaking the skin. When a mother cat has kittens, she will frequently wash them. As they grow older, like all youngsters, they will not want to be washed. They will wriggle and squirm and try to get away from their mother, rather than be washed one more time. As the mother cat has no hands to restrain them, she uses her teeth instead. She will very gently but firmly clamp her teeth around her offspring's neck, and the kitten will immediately become immobile. She will then proceed to wash the kitten, who will probably bear with it for a few seconds, then start to squirm again; so the mother cat will once again clamp her teeth around the kitten's neck until it is still, then start to wash again.

Sometimes a cat will restrain its 'owner' in this way. I say 'owner' in inverted commas as this is the type of owner who is often owned by a cat! In other words, the cat calls the shots and knows he is the boss of the household.

Many owners ignore their cats, except to feed them,

believing they are 'solitary' creatures. They aren't — they're independent but social creatures. Without a rewarding relationship with other cats or their owners, they will become maladjusted. Often a cat treated in this way will believe he is the boss of the household, and will act accordingly, reacting with a restraining bite to stop the owner doing something the cat believes he or she shouldn't be doing (such as moving their legs if the cat is resting on their lap).

Some owners go to the other extreme entirely, doting on their cats and never checking them, no matter what naughtiness the cat is getting up to. Often, these owners feel they will lose their cat's love if they check their behaviour, but the opposite is true.

It is a great responsibility being the group leader or 'boss cat' and your cat would much prefer that this responsibility was shouldered by you. Bring your cat up to know what is and what is not acceptable behaviour and tell him by a sharp word when he's misbehaving. Reinforce this by another sharp word if he does the same thing again; most cats will learn what is acceptable after just two or three cross words from an owner. *Don't* smack a cat to get your message across; you'll just make him nervous.

By laying down *your* guidelines as to what constitutes acceptable behaviour, and by not spoiling your cat by allowing him to behave in an anti-social manner, you will get across the point that you are the leader and you make the rules. A cat brought up in this manner will rarely, if ever, give a restraining bite.

Of course, not all humans in the household fit into the hierarchy in the same position. In my household, I am accepted unquestioningly by all four cats as boss cat.

But my senior cat, Morgan, is very assertive, and she thinks she's my second-in-command! So, although she will never use her teeth on me, she will restrain my nine-year-old son with her teeth if she is sitting on his lap and he fidgets! Fortunately, my son is acknowledged as third-in-command by the other cats, so Morgan is the only cat who will 'bite' him in this way.

Another type of 'aggression' is the *ambush* – when a cat will lie in wait for an owner to pass, then jump out and bite or scratch. This is really just over-active play – although knowing that your cat is playing isn't likely to minimise the pain! The only way to deal with this is to carry a number of small toys around with you in your pockets, perhaps a catnip mouse or a ball. As your cat leaps out, quickly drop the toy and his playfulness should redirect itself towards the mouse or ball.

It would probably also be helpful to spend some time each day in structured play with a cat which behaves in this way. Make it a regular feature, at the same time each day, not too near mealtimes. In the middle of the day would probably be best and ten or fifteen minutes can be spent daily, using up this cat's energy and allowing him free range for his playfulness.

Trail a string for him to chase (keep firm hold of one end of it, as, if swallowed, string can be very dangerous and may require an operation to remove). Throw a ball for him, or a piece of paper twisted into a 'stick' (make sure it is too big to swallow), which many cats will learn to retrieve. Wind-up toys (with no small pieces to break off) are often a favourite plaything which will help to save *your* energy, and a 'fishing-pole' type of toy will help the immobile (or lazy) owner to exercise their cat without moving from a chair. Make one by tying a piece of stout (non-fraying) string to a rod or cane. Tie a piece of tough fabric to the other end and your cat will play with this and chase it. At the end of each play session, with a bit of luck he'll be too tired to chase *you*.

If your cat's aggression falls into none of these categories, do take him to the vet for a thorough check-up. Many cats will become bad-tempered when ill but may have no other symptoms that you have noticed. If they're feeling generally unwell, because of a urological problem, for example, they will be cross and may take out their tempers on their owners. Have the problem treated, and the normal sunny temper of your cat will return.

It is worth trying a course of 'nerve tablets' for an aggressive cat. The B vitamins are generally considered to be the anti-stress vitamins and some bad-tempered humans find their tempers improve once on a course! Buy some ordinary brewers' yeast tablets for your cat — they're very inexpensive. Most cats will love them and simply chew them up as if they were treats. You can safely give two or three tablets per day. It is unlikely that you will give your cat an overdose, as the B vitamins are water-soluble and simply 'flush through' the body if they are taken in too-large quantities. However, don't take any chances and restrict your cat to just a few tablets daily. If these have no effect within a week or so, many pet shops stock a range of herbal tablets, including a herbal nerve tonic tablet which might help in your particular case. Being a completely natural, herbal product, there is no chance that you will do your cat any harm by trying it.

In a few cases, a cat can become aggressive due to a

hormonal imbalance and this can be rectified by a course of tablets prescribed by a vet.

In some cases, vets will prescribe specially formulated tranquillisers for an aggressive cat. This may be the only choice in certain, rare cases. It should be considered only as a very last resort, when you and your vet have investigated every other possibility for your cat's behaviour.

Unfortunately, some vets are now treating pets in the way that certain doctors are accused of treating their patients – by reaching for a prescription pad instead of trying to discover the underlying cause of an unacceptable behaviour pattern. Many vets have little knowledge of behavioural problems and the veterinary associations are only now beginning to urge vets to acquire this knowledge.

On a recent television programme, a 'TV vet' was asked a question about a biting cat and, without asking for any further information, immediately recommended that the cat be put on a course of tranquillisers.

So if your vet seems to be too quick in recommending tranquillisers for any problem, insist on discussing it with him and asking if there are any alternatives which can be tried first. Once everything else has been tried, and there is nothing left but tranquillisers, ask for the lowest recommended dose over the shortest recommended time.

Tranquillisers *do* have a very useful place in veterinary medicine. Aggression will often cease after a course of this treatment, even when the course is finished. But all tranquillisers should be used with care.

Common Problems
AGGRESSION TOWARDS ANOTHER CAT

How to introduce a new cat or kitten

How to introduce a cat to another pet

Reasons for aggression:

- territorial disputes
- cat is still entire
- illness
- return after absence
- allergic reaction
- hormonal imbalance
- stress

Aggression towards another cat

Squabbles between cats in the same household are not uncommon and not surprising, as you'll understand if you put yourself in your cat's position for a moment.

Imagine yourself living peacefully in a household when, one day, a perfect stranger is brought in. He may be bigger than you, he may make it perfectly plain from the outset that he expects to be boss in *your* house, and he may have bad breath and unpleasant eating habits. Yet, without even being introduced, you are expected to share your dinner with him, share your bathroom with him and even share a bed with him! Add to this the fact that everyone else in the household is fussing over this new arrival and completely ignoring you. Wouldn't you feel like punching him on the nose when no one was looking?

This is exactly the position many cats find themselves in when another cat is added to the household.

Harmony in the household goes further back than this, however. If you already have a cat and you are giving a home to another cat or a kitten, try to choose one whose personality will fit in with that of your other cat. If your cat is shy and timid, don't bring home an over-confident bruiser. If your cat likes to be the boss, don't bring home *another* boss-cat; try to find a cat or kitten with a non-assertive disposition.

First introductions are vital, and should be done

slowly. If the new cat or kitten comes from a dubious or unknown home, it is probably best to isolate him from your existing cat for a few days anyway in case he's harbouring an infection. Confine him to one room at first, with food, a clean litter tray and somewhere comfortable to sleep, until a vet has checked him over and passed him as fit.

A new cat won't mind this treatment; he'll be wary and, in fact, will feel more secure getting to know one room rather than a complete home all at once. While he is in there, he will be able to smell the existing cat, who will be able to smell him, so giving them time to get used to one another's scents before they are introduced.

After a few days have elapsed, and you are certain that the new cat is not suffering from any illness, the two cats can be introduced – or, if the cats have not been isolated, start from this point.

Your existing cat should be sitting comfortably on his favourite person's lap and should be having a fuss made of him, with perhaps a few treats given. Another person, preferably not someone living in the house, should quietly bring in the new cat and set him down in the same room, not too near the other cat (who should not be restrained in any way).

Ensure beforehand that the room selected for the introduction has plenty of hiding places. If it doesn't, make a few temporary ones by bringing in additional furniture or leaving a number of large cardboard boxes in the room.

If the two cats have nowhere to hide from one another, they will have an eyeball-to-eyeball confrontation. A cat feels threatened when stared at by another cat as this is a preliminary to a leap and possible fight. With places to hide in the room, no staring confrontations should occur.

It is very important not to allow one cat to run straight at another, especially in the early stages of introduction. This will frighten the cat which is being charged. Some cats, having been on the receiving end of this behaviour, will regard the other cat with trepidation for the rest of their lives.

It is unlikely to happen when your 'introduction room' has been probably prepared, with plenty of furniture or cardboard boxes scattered throughout it. If there is enough clutter, there won't be space for one cat to run at the other.

The new cat will probably take just a few steps before the other cat jumps down to see what is going on, sniffing the stranger and possibly growling. It is important not to interfere; your existing cat must make friends with the newcomer before he sees *you* making friends with the newcomer.

It is almost inevitable that there will be spitting and hissing (which can sound quite bloodcurdling) at first, which may be accompanied by flashing claws, bared teeth – even fighting. This is the cats behaving as nature intended; one cat is telling the other (and proving) that *he* is the boss and this is *his* house. Having once introduced them, do restrain yourself from interfering unless one of the cats is really being injured. This is highly unlikely, as the cats are simply sorting out who will be the dominant one; they're not actually intending to hurt one another. Trim their claws before they meet and they will be less likely to hurt one another by accident.

Only if the fighting has gone on non-stop for an hour or so, or if one cat is actually bleeding, or has become really frightened (gasping, panting with his mouth open) should you then consider separating them, putting the new one back in his room, and trying again the following day.

If you have no success the next day, separate them again for a day. Before bringing them together again, you can help them to accept one another more easily by making them smell the same. Cats rely heavily on their sense of smell and won't like a strange-smelling newcomer. You can wipe over their coats with a solution of one tablespoon of cider vinegar in two pints of water, which will make them smell alike to one another. Be careful not to get the solution into the cat's eyes or over cuts or scratches. If this still doesn't work, rub baby powder into their coats (without getting it in their eyes) and brush out thoroughly.

If it still isn't working, buy or borrow a large cat pen and put the new cat inside it in the middle of the living-room. He should have a litter tray, food and drink and his bed in there with him. The two cats will be able to see and smell one another but not do one another any harm.

The best time to let the cat out again is at feeding time, when both cats will have other things on their minds – and they should, of course, have separate feeding bowls spaced several feet apart.

Once the cats have grudgingly accepted one another, you can help them settle down with a play session in which they're both involved. Trail string for them, or roll a ball and they'll be too interested in what is going on to argue with one another.

A couple of pinches of catnip sprinkled on the floor can help to ease sticky moments as it will help those cats which are susceptible to it to relax. They should sniff it, possibly eat it and roll around in it, which will help take their minds off the other cat.

Catnip (also called catmint) is a herb which is available from herbalists and health food stores and is a useful tool when you occasionally want to help your cat to relax. It does, however, work on the cat's nervous system and dull his senses, so don't allow a cat which has been using catnip outside for at least half an hour afterwards.

Also remember in the early stages to minimise confrontations of all sorts as much as possible. If you are, for example, letting one of the cats outside while the other is waiting to come in, they can come face to face at the door, which is almost certain to lead to argument. If you can, pick up the cat which is waiting to get out and open the door wide so the incoming cat can see that no one is waiting to ambush him. When he comes in, set down the cat you have picked up just inside the door, allowing him to go out in his own time.

If you don't go to these lengths, at least open the door as wide as possible to allow the cats to pass one another without encroaching on one another's space.

It may take some cats a very long time to settle down with one another, so don't expect harmony overnight. Two or three months or more can be needed by some cats to adjust to a new kitten and it has been known for adult cats to take as long as *a year* to adjust to one another. And some cats, although they may settle down to some degree, may never really like one another, even if they are related.

But, if you do want a second (or third or fourth) cat, it *is* worth persevering. The cats will be company for one another; an older cat will be rejuvenated by a younger cat – and you will have twice as much fun! And you may be one of the lucky ones and find that your cats settle down together almost immediately.

If you own a Siamese or other foreign type of cat, it is almost essential to keep more than one cat, especially if you are out of the house much. These cats find companionship essential – either human or feline – and can become quite maladjusted if their capacity for play isn't completely fulfilled.

Should you be introducing a new cat or kitten to a dog (instead of another cat), most of the foregoing will still hold true. It seems to help successful introductions if the two are separated for a few days at first, so they can get used to one another's scent through a closed door. Do keep hold of the dog's collar when first introductions are being made, and don't leave the two alone together until you have been able to observe them together for several days and are sure that a fight won't result. It seems that cats and dogs get along better together when one of them (if not both) is a juvenile when introduced.

Don't try introducing your cat to smaller animals such as mice, gerbils or lizards. Although a rare feline will live closely with small pets without harm, it really is expecting too much of a cat to behave when their natural prey is introduced to them 'on the hoof'. In my opinion, it isn't really fair on either animal to keep something like a mouse in the same household as a cat.

When you have two or more cats in one household, it

is important that each one should have his *own territory* within the house. Keep an eye on both cats to ensure that each is getting his fair share of food – some will eat more and faster than others. They should have their own bowls, but if this doesn't stop food-stealing, they should be fed in separate rooms. They should also have separate litter trays, if used, and beds of their own. It might be helpful to keep the peace if the beds are in separate rooms; that way each cat will at least have peace and quiet at night.

If this degree of separation doesn't minimise fights, you should perhaps consider buying an outdoor cat run, in which one of the cats can spend his days, coming back into the house at night to sleep in his separate room. If this is the only solution, do remember to give him plenty of love and attention while he is in his run – his life will become boring and lonely if he's simply locked away and forgotten.

Do also make sure that feuding cats do receive the same amount of attention from their human family. Cats can become jealous if they believe that someone else is receiving more attention than they are.

I've mentioned earlier that all pet cats really should be *neutered*. If you have one neutered cat and one which is entire, it is almost certain that the entire cat (which will occupy a higher position in the cat hierarchy) will fight with the neutered cat to prove his superiority. If he is a male, he will also get into fights over females with other cats in the neighbourhood. These fights can be quite serious, necessitating veterinary treatment, so it is really worthwhile having him neutered for that reason, if no other.

Some cats which have got along well together for

some time may suddenly start to fight. It's worth taking them to your vet for a check-up, as an *illness* might be the underlying cause. A cat may not be exhibiting any other symptoms which you have noticed but, if his temper suddenly becomes short, pain or discomfort could be the reason.

Cats which have settled down well together may start to scrap again if one of them has been *away from home* for a day or two, for example at the vet's. This is because the cat which has been away will now smell strange to the other cat. They will usually settle down again together within a few days but you may have to keep an eye on them until they do.

If illness or absence is not the cause of aggressive behaviour, you could try questioning your cat's *environment*. Cats will sometimes react in an aggressive way to toxins in their environment or possibly to additives in their food. This is a type of allergic reaction which could be caused by new paint or other chemical treatment to their home, so ask yourself if this could be a cause of an unexpected burst of aggression. You could try changing your cat's diet for a while to see if matters improve. I have known a few cases where a cat will start behaving aggressively for no apparent reason, yet his aggressive behaviour will cease when his diet is changed. It may be that some cats have a type of allergic reaction to some foods, or the additives in them, in the same way that it has been discovered that some humans react aggressively or antisocially due to an allergic reaction to additives in processed food.

Your cat may have a *hormonal imbalance*, in which case you will have to consult your vet. Or the *stress* can be taken out of the situation in some cases by increasing his vitamin B intake or by trying herbal 'nerve tonic' tablets as detailed in the previous chapter.

If everything possible has been tried and your cats still react badly to one another, there are only two final alternatives. One of them can be found a good home elsewhere, or a vet can be asked to prescribe a mild tranquilliser.

Common Problems   STRESS

The stressed cat

- how do cats suffer from stress?
- what are the reasons?
- the effects of stress

How to help your cat unwind

- aerobics
- diet
- herbal remedies
- massage

Stress

Although some of the effects of stress have already been described, it is worth devoting a separate section to the problem, if only because most people simply don't realise that a cat can suffer from stress. You could be forgiven for thinking that the cat, with its fluid, sinuous body which seems totally lacking in tension, is incapable of suffering from this ailment of the twentieth century.

The reason our cats are leading more and more stress-filled lives becomes apparent if we study the reasons why many people own cats.

As we live in an increasingly urban environment, the cat is seen as an 'easier' pet than a dog – it doesn't require walks and there is not the problem of the owner having to dispose of its faeces. Also, as more members of each family go out to work, the cat is seen as the pet which can 'look after itself' all day.

Although the cat is supremely adaptable to all situations, urban environments are not the cat's natural habitat. This territorial creature can feel threatened if he finds himself living in a high-density situation with many other cats, as often happens in cities. He may have little or no access to trees, grass or herbs which he would use as natural markers or herbal remedies when feeling uptight.

A cat is also a sociable creature. He doesn't enjoy

being left on his own all day while owners work — especially as so many owners will leave him without any toys or playthings with which to while away the hours. A bored cat can easily become an unhappy — or destructive — cat.

It is also now more usual for people to move house with regularity. The conservative cat hates having this disruption in his life with its subsequent change of territory.

Families alter frequently too. When there is a divorce or death in the family, few owners stop to think of the effect of the separation on their pet — yet many cats can be devastated if someone they love goes away. An addition to the family — a new spouse, baby or pet — can have the same effect. If a lot of attention is suddenly focussed on someone new, a cat can become jealous and unhappy.

There are innumerable other causes of stress in a cat's life — from an injury to the cat to the owner moving furniture around — all of which can have the same effects.

These are some of the causes of stress, but what are the effects? They can be so wide-ranging that the owner will often not realise that the pet's problem is stress-related. They can include soiling or spraying, aggression, jealousy, depression and destructiveness. The cat may be off his food and seem generally unhappy or unwell.

So what is the answer to stress? Obviously, the ideal solution is to remove whatever factor is causing the stress. However, this is often not possible, so steps should be taken to alleviate the stress as much as possible and to help the cat relax.

Just as a human will unwind with a workout at the gym, a soothing massage or a dry martini, the same principles can be put to work in helping your cat relax.

You can help your cat to unwind by playing with him daily. To most people, playing with their cat is a foreign concept. Although they will happily exercise a dog by throwing a stick, they believe that a cat is capable of exercising himself. Few creatures are as active as a kitten but, as that kitten becomes older, he often becomes lazier and will play less and less – unless he has someone to play with. Play in adulthood will not only keep him healthier by toning his muscles and strengthening his cardiovascular system, but it will strengthen the important bond between cat and owner. Play with your cat regularly and you will find that it will enrich your relationship to a degree you would not have thought possible.

If you can, keep the 'exercise period' to a certain time each day, lasting about ten to twenty minutes, and starting and finishing with a gentle warm-up and cooldown. Roll a ball for your cat to chase, trail a string or large feather for him to stalk, teach him to retrieve a paper 'stick', or invent whatever games keep your cat moving and enjoying himself.

An exercise regime will particularly help a cat which is bored, very young, elderly, is restricted to indoors, or is one of the very active breeds such as the Siamese.

An excellent diet is vitally important to a stressed cat. It must be well-balanced, containing all the nutrients necessary for full health, and can include a supplement such as brewers' yeast (see the section on Food and Drink, page 77).

The equivalent of the businessperson's relaxing

drink at the end of a hard day can also be used to help a cat unwind. Obviously, *don't* give him alcohol, but you could consider sprinkling a little catnip on the floor to help him relax. Catnip is a herb which is available from herbalists and pet stores and is capable of giving those cats susceptible to it (which is most cats) a 'high' which is relaxing but not addictive.

Cats will roll in it, sniff it and chew on it and have a very pleasurable half-hour. Some owners won't give their cats catnip, believing that their cats' subsequent contortions are degrading to an elegant creature, but this is obviously up to individual owners. One word of warning, however: don't allow your cat out of doors for at least half an hour after using catnip, as he will still be under its effect and may not react quickly enough to a dangerous situation.

Just sitting talking to your cat and quietly stroking him will also help him to relax. It is amazing how many people – because of their own hectic schedules – never find time to sit down with their cats. Then they wonder why their cats become aloof and ignore them. If they did find time to spend with their cats, they would find it would also help *them* to relax, lowering their blood pressure and heart rate and making them feel much less pressured.

An extension of stroking is massaging your cat. This not only relieves stress (yours *and* his), but it can also actually help to relieve the effects of some ailments such as cystitis, arthritis and kidney problems.

When massaging a cat, never use oil, and remember always to stroke in the direction of the fur. Don't touch the backbone or windpipe when massaging.

Place your cat on your lap and slowly stroke him

from head to tail until he begins to relax. Then with the flat part of the fingertips, gently massage along his sides, using small circular movements, from neck to tail. Then, again using circular movements, massage the back of the neck and stroke the front of the neck (missing the windpipe) with vertical movements. Then, one at a time, gently massage the legs in a downward motion and gently massage between each paw pad, jiggling each toe. Finish by using circular movements on the tummy and then gently caressing all over.

Anything you can do to reinforce that magic bond between owner and cat will not be only valuable in your relationship but will be invaluable in terms of health and well-being.

Questions and Answers

The following questions are those I am most often asked by cat owners, and are intended as a quick reference. Some of the problems covered are dealt with in more depth in earlier parts of the book, so, where necessary, the answers will be cross-referenced to earlier chapters.

As you will already have seen, many problems have several possible solutions. Where this is the case, you should decide which is most likely to be effective in your case and try that solution first. If, after several weeks, no improvement is seen, try another possible solution . . . and so on. One prerequisite of caring cat ownership is unlimited patience!

Grooming

I have a longhaired cat whose fur has become matted and tangled. She hates being brushed, so how can I get rid of the tangles?

Once a longhaired cat's fur has become really matted, it is practically impossible to comb or brush out the tangles. If your cat will allow you to do so, you can try untangling the knots by gently and carefully using a device designed to unpick sewing stitches. This will take a long time and patience from you both.

If your cat won't allow this, the only answer is to take her to your vet, who will have to cut off the matts, possibly under anaesthetic. Although your cat will have bald patches after this treatment, the hair will grow back again.

If you are buying a longhaired cat, it is important to realise that it will probably be necessary to comb or brush your cat every single day without fail. Owners of longhaired Persians will brush their cats for as much as twenty minutes a day.

Longhaired cats will often shed a great deal of fur, particularly during the moulting season.

So do think twice before giving a home to a longhaired cat. They're often very beautiful – but they usually involve more work than other cats.

Spraying

Some months ago, my neutered cat broke his leg and had to be kept indoors until he was well again. Since the accident, he has been spraying inside and outside the house constantly. I have taken him to several vets and he has been given injections and hormone tablets to no avail. What can I do to stop this behaviour?

When cats spray indoors, it is usually for one of three reasons. First, it can be hormonal. Neutering will often stop this behaviour, unless the neutering has not been completely carried out.

Second, the reason can be territorial. A cat will mark his territory by spraying (or fouling).

Neither of these would seem to be the reason in the case of your cat, as you have sought veterinary advice and your cat is neutered and lives in a single-cat household.

The third reason for spraying (or fouling) the house can be as a reaction to stress, of many different types, and this seems to be the case here.

Your cat obviously found breaking his leg a very stressful experience and he has reacted to it in this way. Don't chastise him – this will only lead to further stress as your cat certainly won't think he's doing anything wrong.

All you can do is use a great deal of patience. Talk to your cat as much as possible, petting him and stroking him, and giving him plenty of love and support. The behaviour should cease as soon as he settles down again and feels more secure (although this may take some months).

In the meantime, follow him around, thoroughly cleaning up after him. If you don't get rid of all the smell, he will be triggered to spray the same area again

Spraying and fouling are dealt with in more detail in the *Common Problems* section, starting on page 49.

Soiling

I have five cats which were house-trained but recently they have started fouling inside the house. One cat seems to be the ringleader. If I get rid of him, will the others revert to normal behaviour?

Cats will foul inside the house for a number of reasons. The most common is probably illness, so it is very important to take a fouling cat to the vet straight away. Once the cat is well again, the behaviour ceases.

However, I think the reason in this instance may be overcrowding. Once four or five cats are brought together in one household, problems such as this can occur (although this isn't inevitable).

If you are determined to 'get rid of' the ringleader (by which I hope you mean finding him a good alternative home) you will probably find he will cease fouling if he is the only cat, or one of two cats, in his new home. He should no longer feel the need to mark out his territory in this way in a household with fewer cats.

As to the others 'reverting to normal behaviour' – they are behaving 'normally' at the moment in the abnormal circumstances in which they find themselves. Large numbers of cats living in one household are more likely to behave in this way, so, if an owner does not want the possibility of fouling (or spraying) to arise, he or she should restrict themselves to one or two cats.

In this case, removing the ringleader *may* mean that the other cats will stop fouling. However, they may not, as the behaviour may have become established and they are still relatively crowded for space.

It might help if the cats can be given certain areas or territories within the house for their 'own', for example, two cats might share one room at night while the remainder share a separate room.

Four or five cats in a household should not be expected to share one litter tray – two or three trays should be provided which should be cleaned out as soon as possible after use.

Fouling and spraying are dealt with in more detail in the *Common Problems* section, starting on page 49.

Litter lover

*I have a kitten who insists on using his litter tray.
Even when he's outdoors, he will come inside to use
his tray. How do I stop him doing this?*

If you can afford the changes of litter, and don't mind
cleaning out trays, it's not a bad thing for a cat always to
use a tray. A change in a cat's waste products is often
the first indication of illness, a tummy upset, worms or
many other problems, and you can't keep track of this if
he always relieves himself in the garden.

However, if you would really prefer him to use the
soil in the garden, start moving the litter tray a little
nearer the back door each day, until it is outside. A
sprinkle of soil on top of the litter will help to give him
the idea.

Some cats won't use the garden because there is no
suitable spot for them. If your garden consists of hard-
baked earth or lawn, you will have to dig up a small
piece as your cat will prefer to use freshly turned soil.
Some cats also feel too vulnerable to relieve themselves outside – they may fear attack when in no
position to retaliate! In this case, put a large box or tea
chest on its side with the litter tray inside. If the cat
then uses this, up-end the box on the toilet area you
have prepared, making sure you cut a hole in one side
for access.

However, if your kitten continues to use his litter
tray, console yourself with the thought that at least he
is not making any enemies among your neighbours,
many of whom intensely dislike (for obvious reasons)
cats using their gardens as toilets.

Plant-eating
How can I stop my cat eating my houseplants?

All cats need some green matter as an aid to digestion. Free-living cats will have 10 per cent or more vegetable matter in their stomach contents, yet many pet cats have little access to grass or other growing things, so resort to eating houseplants instead. This can be very dangerous, as some houseplants are poisonous to cats, including dieffenbachias and poinsettias.

Grow a little tub of grass for your cat indoors so that he has 24-hours-a-day access to green material and he should stop nibbling the plants. Cocksfoot grass is probably the best kind to grow and, if your local pet shop doesn't stock ready-sown tubs, they should be able to order it for you. Be wary of growing it yourself from seed as some grass seed is treated with chemicals which could be harmful for your cat. You can always dig up a clump from the garden but you will run the risk of bringing disease into the house if a neighbour's cat uses your garden.

When you've supplied your cat with his tub of grass, he can then be discouraged from nibbling your houseplants by sprinkling pepper or dabbing diluted Tabasco sauce on the plant leaves.

Wool-eating
My Siamese cat eats clothes and furniture! Although he is fed a good diet, he eats material – wool for preference. If clothes are locked away, he eats the furniture and carpets. Now, against all advice, I give him a garment with his food – he eats a mouthful of

food, then a mouthful of wool! Can anything be done to stop this?

This behaviour is not at all rare, especially among the foreign types of cats such as the Siamese. It used to be thought that the behaviour always had psychological origins; that the cat had been weaned too young, or taken away from its mother too early. Now it is believed that the behaviour sometimes has a dietary origin. Some cats which eat wool are doing so simply because there isn't enough fibre in their diet. This can be a particular problem with some pedigree cats which live indoors and don't have the opportunity to supplement their diet naturally, but it can be a problem with indoor/outdoor cats too.

A cat in the wild would eat *all* its prey – bone, fur, feather and gristle. Cats fed on the average meaty diet today simply aren't given the chance to replace the missing fibre and roughage, so some try to find it elsewhere – such as in the wardrobe.

Buy a big bag of bran from the supermarket and mix a large spoonful into your cat's food at every meal. Wholemeal bread, toast or crusts can also be mixed into his food. Try adding chopped-up vegetables to his food too, in small amounts to begin with. Some cats really love some types of vegetables and won't need any persuasion to eat them. And do make sure your cat has access to grass – if he doesn't (or even if he does) grow a small tub for him indoors so that he can chew on it at will to aid digestion.

If the new diet doesn't stop the behaviour, you could try adding a little pure lanolin (available from chemists) to your cat's food. Some authorities believe that cats will eat wool to get a source of lanolin – although 'wool'-eating cats will eat manmade yarns just as happily.

Do ensure that the wool-eating cat isn't just bored. This can be the case, especially with the highly active and intelligent foreign cats, such as the Siamese. If boredom is the cause, you will have to spend more time talking to and playing with your cat – or buy him another foreign cat for company.

If this does not work, some way of keeping your cat away from clothing and soft furnishings must be found, perhaps by locking wardrobes and replacing floor coverings with vinyl, and using seats and chair covers of vinyl. Alternatively, furniture and floors could be shrouded in plastic sheeting while your cat is in the room!

If your cat has a favourite garment which he eats, sprinkle it with pepper or mustard, or dab it with Tabasco sauce or vinegar in order to discourage him. There is also a repellant spray, used by many vets to stop cats and dogs licking treated wounds, called Bitter Apple. Ask your vet if he or she can supply you with some and spray any fabrics your cat might come into contact with.

It is vital that wool-eating behaviour is stopped immediately as major surgery could become necessary if the wool entangles in the intestines.

Clothes-sucking
My cat doesn't eat clothes but he sucks them! It's very unhygienic, so how can I stop this?

If you are sure your cat's diet is properly balanced, yet

he insists on sucking clothes (often wool) you can only hope to discourage him by removing him from the fabric each time. Better still, try rinsing the fabric in something repellent to a cat such as diluted vinegar, or add a few drops of citrus oils or peppermint oil direct to the fabric. Don't rinse your fabrics in bleach – many cats find that very attractive.

It may be suggested to you that the reason for this behaviour is too-early weaning in kittenhood, but this doesn't always follow. I know cats which behave in this way who didn't leave their mothers until they were three months old. It has been suggested that wool-sucking may be due to the cat wanting to absorb the lanolin from the wool. However, as most yarn these days contains little or no wool and consequently no lanolin, I'm not convinced of this theory.

Other cats will suck the sweatier parts of their owners' clothing, such as pyjamas under the arm!

Skin-sucking
My new kitten sucks my arms and neck. Will he grow out of this behaviour?

Yes, he will, if you discourage it – but this may take a year or more. Tell him to stop each time he behaves this way, and remove him each time. Do ensure that he isn't trying to tell you he's hungry or thirsty. If you can wear long-sleeved, high-necked clothing, this will discourage him. If he sucks one particular part of your skin regularly, paint this part with nail-biting solution, bitter aloes or vinegar. You'll need patience but he should, eventually, stop!

Holiday catteries

I'm going on holiday soon and have booked my cat into a cattery. I've inspected the cattery and although it was clean and everyone was so kind, I'm really very worried about how my cat will settle down there. He's never been away from home before and, I have to admit, is quite spoilt. The runs look like prisons and I think he'll hate having cats all around him.

It is vital to check out a cattery personally. A surprising number of people don't bother to do so and simply book their cats in – sometimes with disastrous consequences. If you find one where the boarders look contented and have separate runs and chalets (and are *not* living in boxes in tiers), where the owners appear competent and caring and interested in your particular cat, then book him in and stop worrying.

A cat doesn't see a cattery through human eyes – what looks like a prison to an owner looks like a safe, secure little home to a cat. He won't mind seeing cats on either side of him – he will know they can't get at him and they won't pose a threat. Many cats strike up holiday friendships with neighbouring cats.

Ask the owners if you can supply your cat's own bed, blanket or toys, to make him feel more at home and if he can be fed his usual diet (which most cattery owners will be happy to do).

'Spoilt' cats often enjoy their time in a cattery more than their owners expect – they like being treated as a cat for a change!

Scratching furniture
My cat scratches furniture, carpets and wallpaper.
How can I stop him?

Cats *need* to scratch. It marks their territory, helps them
stretch and exercise, and keeps their claws in good
shape. It is important that each cat has something he is
allowed to scratch. Scratching posts can be purchased
from pet shops or you can make your own by winding a
rope around a sturdy table leg, or attaching a piece of
scrap carpet to a wall. Scratching posts should be at least
30 inches high to allow your cat to stretch.

Having provided him with a post, tell him 'no' very
firmly when he scratches anything else, and remove
him from what he is doing. If he persists, squirt him
with water from a water pistol – he won't connect the
punishment with you but with behaving in a certain
way, and he will soon stop.

Covering furniture or carpet with a sheet of plastic
will stop your cat from scratching, as cats hate the feel
of plastic under their paws.

It helps if you choose your furnishings to suit your
cat! Smooth, silky fabrics are best on furniture as
anything textured or rough will be attractive to a cat.
Hessian wallpaper is a great mistake in a 'catty' household – cats just love to climb up it!

A cat's claws can be trimmed slightly – just taking off
the top – to 'blunt' them a little. This will help the
problem to some extent.

In some countries, cat owners will have their cats
declawed – the claws will be surgically removed by a vet
– to save wear and tear on their furniture. This operation is the equivalent, on a human, of having all your

fingernails removed down to the end of the first knuckle on each finger. As such, it is a disfiguring operation which leaves a cat practically defenceless and unable to grip or climb in a dangerous situation.

It is now also possible to have coloured beads fitted over a cat's claws to prevent scratching. This is unnatural and uncomfortable for the cat and a caring owner will not consider using them.

A properly trained cat will scratch very little but, if you want to live with a cat, you must be prepared to tolerate some scratched furniture sometimes, even if it's your best chair. Your cat will always scratch there because, by doing so, he's marking the chair which smells of *your* scent with *his own* scent. Think of it as a friendly, sociable action and you might not be so cross about the claw marks!

And, of course, some cats will scratch and cause other havoc simply because they are bored. If you can't spend more time with your cat, consider acquiring another cat to keep him company while you're not around.

Hunting
I love cats but I love birds and mice too. How can I stop my cat hunting them?

If you have a cat, it is important that you don't try to attract birds and other animals to the garden by feeding them. It is unfair to expect a cat to sit and watch lots of small creatures in his territory without doing anything about it.

You could try fitting him with a collar which has a small bell attached; it might warn some small animal of an impending attack. However, many cats are expert at sitting perfectly still until the fatal pounce, so this might not be completely effective.

Aversion therapy might be the only thing which will work on a confirmed hunter, although I don't like to practise this on my cats. Tie a bundle of feathers to the end of a piece of string attached to a pole. Have handy a 'loaded' water pistol or a very thoroughly washed out washing-up liquid bottle full of clean water. Dangle the feathers in front of your cat and, when he attacks them, squirt him with water. He won't like this but won't realise you are doing it to him. He will associate it with attacking the feathers and should soon cease doing so.

Fencing

I have just moved to a house on a busy road and am worried about my cats in the traffic. I want them to get plenty of exercise in safety. Is there some way in which I could fence in my garden? And would this be cruel to them?

More and more people are fencing in their gardens to keep their cats safe from traffic, theft, cat fights, infection, or even poisoning. It is an expensive option but very well worth while in many people's opinions.

The fence should be about six feet high, covered in wire mesh. The top of each fence post should have another piece of wood, approximately two feet long, attached and facing inwards at an angle of about 45°. This also should be covered with wire mesh. Even if a cat is able to climb the bottom part of the fencing, he will be unable to climb over the piece which faces inwards. The fencing will have to be continued over shed roofs, and overhanging branches should be lopped off so that your cat can't use these to 'escape'.

It is certainly not cruel to fence in your garden for the safety of your cats. They will still be able to enjoy exercise, fresh air and sunshine, without the chance of coming to any harm from other cats, dogs, humans or vehicles.

Leash-training
I live in a flat without a garden. I don't want my cat to run free as I don't believe he would be safe. How can I exercise him in safety?

You could try leash-training your cat so that he will go for walks with you. Some cats take more readily than others to leash-training, especially the foreign types of cat such as the Siamese.

First, allow your cat to become used to wearing a collar without a leash. Put a collar on him for five or ten minutes each day, perhaps making it a pleasurable experience by giving him a treat to eat at the same time.

You should then allow him to become used to a harness. Special cat harnesses are available from pet shops which fasten around the neck and body of a cat and these are much safer than trying to walk a cat just in a collar, which they can easily slip out of.

After a few days of wearing the harness for ten or fifteen minutes per day, you can attach the leash and walk around with your cat. At first, allow him to walk where he wants to, later rewarding him for walking with you by giving him some more treats.

When he's quite happy with the harness and leash, take him somewhere quiet outdoors to walk, keeping well away from dogs, children and other distractions for the first few visits.

Weight-watching
My cat is becoming very fat! I share a flat with another person and my cat tells both of us that the other one has forgotten to feed him – so is fed over and over again. I also suspect he is pulling the same trick on neighbours and being fed elsewhere as well. How can I keep his weight down?

An overweight cat should always be checked by a vet before being placed on a diet as too much weight in a cat can be a symptom of ill-health. If the vet is satisfied with his health, then cut down food gradually, first of all cutting out all snacks.

When several people live in one household and end up feeding their cat several times over, a cardboard sign should be made and displayed. One side should read 'cat fed – morning' and the reverse should read 'cat fed – evening'. Whoever feeds the cat turns the sign around so that the other person knows it has been done.

Ask neighbours not to feed your cat but, if you suspect this is still happening, feed your cat a food which he is not particularly fond of. He will then only eat at home when he is genuinely hungry.

Garden problem
How can I stop a neighbour's cat from using my garden as his toilet?

If you can't fence off your property so that the neighbour's cat is unable to enter, you could try a purpose-made deterrent available from pet shops or ironmongers.

Or you could try scattering pieces of orange or lemon rind around. Many cats don't like citrus oil and will keep away from it. Seedlings can be protected by small twigs, netting or wire mesh. For additional protection, hang mothballs from twigs or branches. Cats dislike the smell, but these should be used with care as they can be toxic to cats. Of course, if you have a cat of your own, these measures will also deter him!

A kinder way of dealing with the problem is to keep a freshly dug patch of soil or a clean sand box (from which children should be excluded) in one part of the garden. Cats will prefer to use these as toilets (as long as they are kept clean) and will probably not bother with the rest of the garden at all.

Or, blow up a number of balloons and attach them to the area you want protected. They will bob about, discouraging most cats but, if any cat is inclined to investigate further, the noise when they burst will persuade him that there are pleasanter places to be.

And if you do have a cat of your own, he will help to keep neighbours' cats away, as 'his' garden will form part of his territory.

People shy
I bought a kitten which I was told had been raised among his breeder's family. He has made friends with our other cats but he spits and runs from all humans. If I pick him up, he looks petrified. Will he always be like this?

I'm sure this kitten will accept you more as time passes if you treat him with great patience.

It sounds as though he might not have had much human contact in early kittenhood; this is vital if kittens are to build up trusting relationships with people.

Spend regular time with your kitten, just quietly talking to him at first and later, as he gains confidence, gently stroking him as you talk. You might find it helps if you get down to his level and actually lie on the floor while you talk to him – that way, you won't be looming over him.

If he's susceptible to catnip, sprinkle a little on the floor during your sessions – you should find that it will help him to relax.

Don't be too quick in trying to pick him up. Let him make the moves in his own time. He may begin to accept you more as he sees other cats react in a friendly way towards you. He may never be quite as friendly as your other cats, but I'm sure that with time and patience he will come round.

Feral pet

A stray cat has been in our neighbourhood for a while and recently gave birth to kittens. When they were about two months old, I caught all three (it took two weeks!) and brought them indoors. The mother cat obviously was a pet at one time and she's fine with me but her kittens, after two weeks, still won't let me stroke them. Sometimes they will approach me and rub against me but then they run away. They seem quite happy in their new life but I wish they were friendlier.

Actually, you're doing very well indeed if the kittens have got to the stage of rubbing against you within a few weeks! Being born feral (wild) they have never been 'imprinted' with humans, so it will take much longer for them to become used to you than it would for kittens born in a house litter. They are simply frightened of these enormous, strange-smelling, odd-sounding creatures (humans).

All the points in the previous letter apply. It will just take a great deal of time and patience to get to know these kittens, but do persevere in talking quietly to them and spending time with them each day. Do play with the kittens by trailing a string or gently rolling a ball to them, if you can do this without frightening them.

Give them plenty of time and don't even try to stroke them until, by the kittens' attitudes, you realise that they actually *want* you to stroke them. Just don't rush them. Kittens can sometimes come around quite quickly (i.e. within a month or two) but people who have given homes to feral, full-grown cats sometimes find it can take *years* before the cat will accept them. In this, as everything else, cats are individuals. Other full-grown ferals have settled down in homes in a few weeks.

Catnip
Where can I buy catnip and what can I use it for?

Catnip (sometimes called catmint) can be purchased from most herbalists and some health food stores. You can grow your own, from seed, during May and June, although it is a difficult plant to grow until the root system is established. Seedsmen Thompson and Morgan sell the seeds and you should specify that you require *Nepeta Cataria* rather than the ornamental varieties.

Catnip has many uses, from stuffing cat toys to encouraging cats to use their scratching posts or litter trays. It relaxes those cats which are susceptible to it and can make the introduction of a new cat into the household much easier.

However, it is very important to remember that catnip works on the cat's nervous system and dulls his senses. If he has been using catnip – or even just playing with a catnip toy – keep him indoors for at least half an hour afterwards to allow the effects to wear off.

Unbalanced diet

I am the owner of two young cats who are determined to eat only the 'real fish' expensive brand of tinned food. Even when this is mixed with tinned meat food, they pick out the fish, leaving the rest! Apart from the expense, I am worried that they are not getting all the necessary nutrients from a fish-only diet. Is this just a question of who has the strongest willpower?

You are quite correct when you say that your cats would not receive all the necessary nutrients from a fish-only diet. No cat would lead a long and healthy life if he was fed only fresh fish.

However, the expensive brands of 'real fish' cat food are a complete food containing added minerals and vitamins. They contain all the nutrients currently believed to be necessary to maintain a cat in good health.

Some of the cheaper tinned 'fish' foods actually contain very little fish, but do contain other meats and meat by-products.

Your cats will probably be receiving an adequate diet if you continue to feed them tinned food with fish, with some fresh meat added for variety. Try to add other meat products to this to enrich their diet. Try them with lightly cooked liver once a week or once a fortnight – most cats enjoy this. They could have some raw mince (for human consumption – not pet mince) occasionally as a treat. A raw egg yolk (not the white) once a week or so is very nourishing.

In the meantime, you could continue to try to mix a little of another flavour of tinned food with the 'fish' tinned food. Perhaps, to make it more palatable to begin with, you could mix it thoroughly with fish stock.

Your problem points up once again how important it is to ensure that kittens receive a varied diet from an early age.

Bald spots
My cat has a bald spot on his tail where he has been sucking it.

If a cat sucks any part of his anatomy, he should first be taken to the vet, who will check him for injury or infection, which might cause this behaviour. A foreign body, such as a thorn, will be sucked by a cat trying to extract it. Bald spots are sometimes caused by ring-worm, a very contagious fungal growth, which gives the skin a scaly appearance, with raised edges, and a vet will be able to treat this too. Sometimes the cat may be sucking part of his body for comfort; if he's feeling lonely, or unsure of a change in the household. In these cases, comfort and reassurance should help to stop it. Bald spots, if not caused by foreign bodies, infection or sucking, are often caused by fights. Cats seem to be expert in lifting chunks of fur out of one another without even leaving a scratch underneath. If the bald spots are around the head, chest or ears and there is no sign of ringworm or other infection or foreign bodies, they are probably caused by fights.

It is not uncommon for a cat to suffer a bald spot on the chin and again veterinary advice should be sought. It is sometimes caused by an allergic reaction to plastic feeding bowls, so if your cat feeds from a plastic bowl, a switch to ceramic or stainless steel may help the problem clear up. It can also be caused by an allergy to something in the cat's diet, so a change of food may help.

Bad breath

I have a new kitten who is delightful. He washes himself regularly – but he smells! Do I have to bath him?

Apart from the decorative longhaired pedigree cats, it is rarely necessary to bath a cat. They are fastidious creatures and usually manage to keep themselves very clean. Your kitten may be teething, which can sometimes cause bad breath. Then his saliva, distributed around the body by washing, will also smell. In this case, the smell should disappear when the kitten stops teething but, to be sure, it would be best to take him to the vet for a check-up, in case there is some other problem.

Bad breath is sometimes a problem with an older cat and a visit to the vet for teeth-cleaning will often solve the problem. Help to keep your cat's teeth clean by giving him some food which will exercise teeth or gums, such as a *very few* pieces of dry cat food before bed (ensure that drinking water is available too) or raw fresh meat occasionally, served in chunks for your cat to tear apart.

Over-active

My cat is 10 months old. He digs up my carpets and chews off the backing, bites my wrists and ankles (purring while he does so), jumps up and down to switch on and off my lights and copies everything I do. I love him but he's a handful. Will he ever settle down?

This is obviously a very healthy, energetic cat. In human terms, he's a 'teenager', and teenagers can be a bit of a handful with seemingly unlimited energy. He should settle down as he matures – around 15 months or so.

In the meantime, he should be allowed plenty of exercise – indoors and out. A bored cat will sometimes become destructive, chewing carpeting, scratching wallpaper and so on.

Biting his owner is probably also a misplaced play activity – he wants to play but is doing so too roughly.

Spend ten or fifteen minutes each day (at least) playing with him, rolling balls, winding up a toy mouse for him to chase, trailing a string for him to pounce on – even try training him to 'fetch' a piece of rolled-up paper.

Make sure an active cat is receiving a good, well-balanced diet. In a few cases where a cat is behaving in a 'hyperactive' manner, I have found a change of diet sometimes helps. Some cats may react to a manufactured cat food in this way – perhaps because of an allergy to some of the additives – so it's worth trying a new flavour, a new brand, or producing some home-cooked meals to see if this leads to an improvement in behaviour.

Cat-flaps

I had no trouble persuading my cat to go outside using his cat-flap. However, he will not come back in unless I open the door for him.

It is usually easy to train a cat to use a flap by first of all propping it open with the cat on one side and you on the other – holding a plate of food or a few treats. After a few tries, the flap can be lowered and the cat rewarded when he comes through it.

But sometimes a cat won't use a flap – in or out – if he cannot clearly see what is on the other side. Cats always like to have a clear escape route in front of them; they will always pause to look around before going through any flap or door. If they can't do this, they may not use the flap. There are clear plastic flaps on the market which may help with this problem.

Or it's possible that a cat simply doesn't like using a

flap. It may have been installed at the wrong height – it should be about 6 inches from floor level. He may not like having to push it open with his head, or he may have trapped his paw in it one day when you weren't around and this could be putting him off. If buying a cat-flap, test it in the shop by pushing it open with one finger and then withdrawing your finger. If it traps your finger, it could trap your cat's paw.

I prefer not to use cat-flaps at all. Not only does this stop neighbours' cats coming in and making messes, or fighting with my cats, but it also means that I can keep track of my cats and know exactly which ones are inside and which ones are out at any time.

If you do decide to install a cat-flap, remember to buy one which locks – essential to keep your cats in at night and the neighbours' cats out. Remember, too, to install it at more than arms' length from the door lock. Burglars sometimes break into a house by putting their arms through the cat flap and unlocking the door.

A new kitten
I have a very possessive cat. Several times I have tried to introduce a new kitten into my household but each time my cat has been jealous, refusing to eat, staying out or hissing at the new kitten. Each time I gave him five weeks to settle down but he didn't so I had to rehome the kittens. I'd still like a kitten – should I try again?

It can take three months or even more for two cats to settle down together, so anyone thinking of introducing a new cat or kitten into their home should be willing to make this investment in time and patience! Occasionally, two cats will never be particularly happy together and the owner may find that he or she is spending time and effort keeping them apart to some extent.

See the section in *Common Problems* on page 97, for ways of introducing a new cat into a household. It may seem a lot of bother, but will probably save more bother in the long run and, if you are not prepared to do it, perhaps you should think again about having another cat.

With the very possessive cat, the owner should try to visit the new cat or kitten many times before bringing him home, ensuring that the smell of the newcomer is on the owner and his or her present cat becomes used to the smell before he meets the newcomer. Build up the existing cat with a good diet before introductions are made – he will be heavily stressed by the 'usurper'. Sprinkle a little catnip around to help him relax and have a stiff drink yourself! If you are anxious or nervous about the introduction, this will communicate itself to your cat, who may react aggressively. And tell your cat beforehand that another cat will be coming to live with him (but that he will still be your favourite). You may feel foolish talking to your cat, but they do seem to understand a surprising amount!

No affection
I adopted a stray cat many years ago and I love him dearly – but he never shows me any affection. In fact, when I speak to him he turns his back on me!

Have you taken your cat to the vet for a check-up? Unfriendly behaviour may show that your cat isn't feeling completely well. In fact, all cats should receive a veterinary check-up at least once a year (perhaps at the time they receive their cat 'flu and enteritis booster injections) and this is even more important once a cat reaches the age of seven or eight and is no longer 'young'.

If he is passed as fully fit, is he receiving all the care and attention he has a right to expect? He should have two nourishing meals a day, drinking water always available, a quiet, comfortable bed indoors (no cat should be left to roam the streets at night), a clean litter tray available at all times, a few toys and a scratching post. He should also have the time and attention of his owner regularly.

Many people believe that, because cats are 'independent', they don't need any company. This is untrue, as the cat is a social creature and you cannot expect a cat to be friendly towards you if you spend most of your time ignoring him. Everyone should put aside some time each day to be spent with their cat; talking quietly to him if he is an inactive or elderly cat or playing games with him if he is young and energetic.

Most cats will begin to respond if regular time is spent with them each day, and will even look forward to that part of their day.

With a nervous, or ex-stray cat, it could help if the owner gets down to floor level when conversing – that way the cat won't feel threatened by an enormous creature towering over him. Speak quietly too; the cat turning his back on his owner is adopting a non-aggressive stance – he's trying to make himself as inconspicuous as possible. In this case, he might still be a little frightened of his owner and patience and quiet conversation over a long period should help to allay his fears.

Moving house

I'm moving house soon and am worried about losing my cat when I move. How soon should I let him out? Should I put butter on his paws?

Moving house is traumatic for cat and owner alike. You can make it easier for your cat by putting him in a quiet room where he will be undisturbed (perhaps the bathroom) while the removal men pack. Give him food, water and a litter tray, and perhaps a toy to play with or some catnip to relax him. When the removal men have gone, put him in a secure cat-carrier and take him to the new house. Do the same thing again – put him in a quiet room until the removal men have unpacked and gone. Prepare a hiding place for him in the room you have chosen – a cardboard box with the lid taped shut and a cat-sized hole cut in the side is ideal. He will be feeling insecure and will want a bolt-hole – it's much easier to make him one than coax him out from under the bath or from up the chimney. Block off all escape routes, close doors and windows before allowing your cat into the rest of the house to explore.

I believe he should be kept indoors for at least a week, but this varies from cat to cat. A very territorial cat will be upset by a move and may take a month to settle down – you'll have to judge for yourself when your cat has settled well, then let him out. Accompany him on his first trips outside and keep the outings short – ten minutes or so. Call him in with you when you go – he'll probably be glad of an excuse to go back to familiar territory.

Butter on the paws will help – by licking it off, he will cover his paws with saliva which will then be distributed around the house, making it smell more familiar to him.

Homing instinct
Since moving house some months ago, our cat has returned to our old house at night every few nights and seems unable to find his way back to us again. What can I do? I'm very much against a freedom-loving cat being restricted in the house.

The answer here is obvious – keep your cat locked in at night. This cat obviously had too much 'freedom' in his new home too soon – before he had put down roots and accepted it as his new territory. Now he is given the 'freedom' to go wandering off at night, running the risk of being lost, stolen or run over. Although cats love to explore and run around in the fresh air, they don't usually need nearly as much freedom as they're given! Watch the average cat spending a day in the garden – at least two-thirds of his time will be spent lazing about or sleeping. A cat allowed 'freedom' at night will spend most of the night either looking for somewhere comfortable to sleep, or else sleeping.

A cat trained to spend his nights indoors from kitten-hood will not even ask to be allowed out. An older cat will complain at first, if he has been used to going out and is then stopped from doing so. But to stop the possibility of harm coming to your cat, it's worth persevering. Close him in a secure room each night – one where you can't hear him complain if this bothers you. Make sure he can't ruin your carpet by scratching (the kitchen may be suitable) and make the room comfortable for him with a litter tray, food and water. Feed him just before bedtime and that will help him to settle for the evening. Then close the door – and ignore him until morning.

Don't rush to him because he miaows plaintively – cats are wonderful actors and if you weaken you'll never train him. As long as you are sure he is safe and secure, *ignore him*. Be reassured by the fact that you are doing the best you can for him and he will soon settle down.

Vegetarian diet

I'm a vegetarian and for several years now I've been feeding my cats on a vegetarian diet with no ill effects. I've recently been told that cats shouldn't be fed on a vegetarian diet but I don't believe this.

Cats have an absolute requirement for taurine in their diet. This is available to them *only* in meat, fish and shellfish. Taurine is an amino acid which is essential to cats to keep their eyes healthy. Without taurine in the diet, a cat will suffer retinal atrophy and will go blind. This may take several years but it is inevitable. You must change your cats' diets immediately to a good-quality cat food or to fresh meat and fish. Any vegetarians whose beliefs preclude them from feeding their cats meat must find their cats a good alternative home where the animals will receive the diet they require.

Dog food does not contain taurine, as the dog is able to manufacture this itself in its own body. So, again, a cat fed nothing but dog food will not remain healthy.

Nervous cat

My cat is extremely nervous and unhappy with my other cats. When allowed out, he goes away and hides all day until I persuade him to come home. He fights with my cats and runs and hides from strangers. What can I do to make him happier?

First of all, take him to your vet for a thorough check-up to see if there is a veterinary reason for his behaviour. He may be suffering from an illness or infection which is not immediately apparent but which is affecting his behaviour.

If the vet says he is completely well, then it's a case of trying one thing at a time. He may have suffered from some trauma (being chased by a dog or hurt by a human) of which you are not aware, so only patience and kindness over a long period may help him. Play with him gently and spend as much time as you can; sprinkle catnip on the floor or in his bed to relax him further.

If the other cats in the household are causing problems, try separating them – at least at night. Place his bed high up, on a stable piece of furniture – he'll feel more secure at a height.

Feed him well to help combat stress. Make sure his

diet is good and contains all the nutrients he needs. Give him extra B vitamins – the anti-stress vitamins – in the form of Brewers' Yeast tablets. These tablets are inexpensive and most cats enjoy them so much that they will eat them as if they were treats. It is unlikely that you'll give your cat an overdose of B vitamins as they are water-soluble and will simply flush through the body if given in excess.

Make sure your cat has his own food bowl and litter tray. He may just be a subservient cat which is low down in the 'pecking order', so do ensure that he receives his fair share of everything, including your time.

Herbal 'nerve tonic' pills are available from most pet shops and some owners have found that they help their nervous cats.

If he continues to seem unhappy, take him back to your vet. Sometimes a hormonal imbalance can make a cat nervous and a course of hormone tablets can put the problem right.

If all else fails, and the cat is really miserable, a course of tranquillisers may be prescribed (see *Common Problems* on page 94). You may also consider finding the cat another home, perhaps one where he is the only pet. His problem may simply be a clash of personalities – with your other cats – or with you!

My cat next door
My cat is eight years old and has always been happy and friendly. However, I recently bought a dog and, since he arrived, my cat won't come home. He has moved in with my neighbour and I just can't tempt him back!

Your cat is telling you very plainly that he doesn't like the new addition to the family. It is not uncommon for a cat to 'pack its bags and go' when a new member joins the family, whether it be a dog, another cat, a new baby or even a new spouse. It demonstrates the basic difference between a cat and a dog – if a cat doesn't like its home conditions, it will leave to find a better one!

There is less chance of this happening if introductions are properly made (see 'First Introductions' in the *Common Problems* section on page 97). Make sure you make a fuss of your cat and spend time with him before, during and after any change in family circumstances. Cats are capable of feeling jealous and left out, so you will have to treat him with tact.

Once the cat has actually left home, it is more difficult to persuade him to return. Perhaps you could ask someone to take your dog for a long walk while you bring in your cat, make a fuss of him and give him his favourite meal. You could try doing this daily; at the same time asking your neighbour not to feed him. After a week or so, you could try re-introducing your cat and dog as described previously.

Do bear in mind that your cat is quite middle-aged at the age of eight and is probably set in his ways. It may take a lot of time and patience for you to regain his trust.

If he makes it plain that he's determined to live next door, the only solution may be to ask your neighbour if she's willing to keep him. If your neighbour is a pensioner (and many cats seem to 'adopt' pensioners – perhaps because they have more time to fuss over them), you might ask her to 'foster' your cat. That is, your cat might live next door but you could pay for his food and any medical treatment. That way, you would have the benefit of seeing him regularly and knowing that he's happy and not becoming a financial burden on anyone else.

If your neighbour is unable or unwilling to give your cat a home, you will have to re-double your efforts to persuade him to come back and you might be helped in this if the weather turns wet or cold!

Green-eyed monster
I love my cat dearly, but my husband doesn't. He ignores the cat and doesn't even like being in the same room as him. It causes arguments between us, but what can I do?

This is always a sad situation for everyone concerned. Your cat will be aware of your husband's dislike and will be unhappy – cats can become very depressed if they feel they're not liked, and may even start behaving badly, which makes the situation worse. You are in the middle trying to keep the peace. It is possible that they are jealous of one another and of the attention you give the other.

Many men profess to dislike cats. Very often it is simply a case of not knowing the cat or understanding him – and what isn't understood is often disliked. It might help if you helped your husband to understand feline body language.

Some men believe it isn't 'macho' to like cats (although tough guys Gary Cooper, Edward G. Robinson and Humphrey Bogart were all keen cat-lovers). Such men will tell everyone that they hate cats, but can sometimes be seen giving the object of their 'hatred' a surreptitious stroke. Tell your husband that cat-strok-ing is now a widely recognised therapy which lowers blood pressure and combats stress.

You will have to be a diplomat to ensure that each of them feels equally happy and wanted. But if all else fails, you will have to consider rehoming him. Then, you and your cat can live in peace together!

Index